WALKING W
DINOSAURS
THE EVIDENCE

WALKING WITH DINOSAURS
THE EVIDENCE

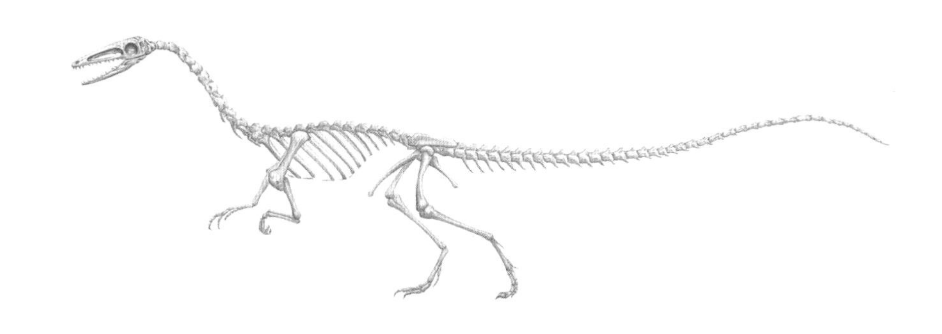

DAVE MARTILL
AND
DARREN NAISH

London, New York, Sydney, Dehli, Paris,
Munich, and Johannesburg

Publisher: Sean Moore
Editorial director: LaVonne Carlson
Project editor: Barbara Minton
Editor: Jane Perlmutter
Art editor: Gus Yoo,
Production director: David Proffit

This book is published to accompany the television series
Walking with Dinosaurs which was produced by the BBC and
first shown on BBC1 in 1999.

Executive producer: John Lynch
Series producer: Tim Haines
Producers: Tim Haines and Jasper James

First published in 2000 by BBC Worldwide Limited,
Woodlands, 80 Wood Lane, London W12 0TT

ISBN: 0-7894-7167-1

Commissioning Editor: Joanne Osborn
Project Editor: Helena Caldon
Art Direction: Linda Blakemore
Book Design: Martin Hendry
Digital images created by Framestore
Stills composited by Alchemy Creative Services

BBC Worldwide would like to thank the following for providing photographs and
for permission to reproduce copyright material. While every effort has been made to
trace and acknowledge copyright holders, we would like to apologize if there
have been any errors or admissions:
Pages 50, 62 and 150 Ardea, London; 79 Insitute und Museum für Geologie und
Paläntologie der Universität Tübingen; 115 and 159 Natural History Museum, London;
127 Professor Pat Vickers-Rich, Monash University, Melbourne and Thomas H. Rich,
Museum of Victoria, Melbourne, Australia; 170 Science Photo Library.

Set in Monotype New Baskerville and Stone Sans
Printed and bound in Great Britain by Butler & Tanner Ltd, Frome
Colour separations by Radstock Reproductions Ltd, Midsomer Norton
Cover printed by Belmont Press Ltd, Northampton

CONTENTS

The Science Behind *Walking with Dinosaurs*

W*ALKING WITH DINOSAURS* is a wildlife documentary set in the Mesozoic Era. Never before has a television series attempted to portray accurately the life habits of dinosaurs as though they were alive and being filmed. *Walking with Dinosaurs* shows prehistoric animals feeding, breeding, killing, and interacting in a way never previously depicted and is a brave attempt to resurrect the dinosaurs and their world.

This novel approach posed innumerable problems. Clearly dinosaurs existed: their fossil remains show this to have been the case. That they walked, fed, bred, and died are inescapable facts if we are to assume that they were once living, breathing animals. Bringing these animals back to life for *Walking with Dinosaurs* relied on two primary sources of information. Paleontologists provided data on the animals' sizes, shapes, diets, and modes of locomotion as inferred from their fossil remains. Geologists provided information on the ages, the environment, and the climates of the ancient worlds in which the dinosaurs lived. In this book we explore this information to show the scientific methodology behind the scenes of the "documentary" *Walking with Dinosaurs.*

The scientists behind *Walking with Dinosaurs*

To resurrect this ancient world, the creators of *Walking with Dinosaurs* collaborated with paleontologists around the world. The BBC consulted a group of eminent scientists as advisors, many of whom have devoted years of research to special areas of Mesozoic paleontology.

Professor Michael Benton of the University of Bristol, England, has worked on all areas of vertebrate paleontology but is best known

◁ The undisputed king of the end Cretaceous world, *Tyrannosaurus rex* eyes its domain.

for his important work on life in the Triassic period, the time when dinosaurs first evolved. Professor Kent Stevens of the University of Oregon, is a computer expert who has recently developed software that allows three-dimensional computer models of animal skeletons to be manipulated in cyberspace. These models have provided new interpretations of the neck structure of the immense long-necked sauropod dinosaurs – data that was crucial to the reconstructions used in *Walking with Dinosaurs.* Dr. Dave Martill of the University of Portsmouth, England, is best known for his work on the fossils of the Oxford Clay, a rock formation that yields spectacular remains of marine animals such as the *Liopleurodon* and ichthyosaurs that are featured in *A Cruel Sea.* Dr. David Unwin of Berlin's Humboldt Museum, works on the fossil record, biology, structure, and biomechanics of pterosaurs, the flying reptiles of the Mesozoic skies. Dr. David Norman of the University of Cambridge, England, has studied the structure and life habits of plant-eating dinosaurs; in particular, he is the world expert on *Iguanodon,* one of the most famous and abundant of all dinosaurs. Dr. Jo Wright of the University of Wyoming worked as a researcher for the entire series; she brought her own skills as an expert on dinosaur footprints. Dr. Thomas Holtz of the University of Maryland is an acknowledged expert on the predatory dinosaurs called theropods and is particularly well known for his studies on the evolution and biomechanics of tyrannosaurs. Dr. Ken Carpenter of the Denver Museum of Natural History, Colorado, has extensively studied stegosaurs and related dinosaurs and is an expert on the Jurassic Morrison Formation.

Many other experts were consulted regarding aspects of animal behavior, biomechanics, vocalizations, and color patterns. The *Walking with Dinosaurs* advisory team was in constant contact via the Internet. Because of the importance of plants in all the scenes depicted in the series, paleobotanists were also consulted and were especially important in ensuring the background scenery accurately reflected ancient landscapes. Modern broad-leafed forests do not resemble the fir and auraucarian (monkey puzzle) forests of the Jurassic, and it was crucial, therefore, that the background was accurate. Fortunately, there are many plant species alive today that are closely related to, and, indeed do, resemble, many of the ancient Mesozoic floras.

The age of dinosaurs: dating the past

It was vitally important that the dinosaurs depicted in *Walking with Dinosaurs* were placed in an accurate time context. The Mesozoic Era, the Age of Dinosaurs, is calculated to have begun 245 million years ago and ended spectacularly 180 million years later in a mass extinction event. During this time, dinosaurs evolved from small, agile, bipedal (two-legged) reptilian ancestors into giant killing machines, plant-eating elephantine monsters and a plethora of other forms. Similarly, over the same period, the seas saw the evolution of diverse marine reptiles, and the skies witnessed the appearance and burgeoning of the pterosaurs, the flying reptiles.

It was crucial that the animals depicted in *Walking with Dinosaurs* were at least roughly contemporary, but it was necessary to condense time a little. For example, the giant *Liopleurodon* from the Middle Jurassic is portrayed with the long-tailed pterosaur *Rhamphorhynchus,* known from the slightly younger Solnhofen Limestone. But scraps of pterosaur bone that are remarkably similar to those of *Rhamphorhynchus* have been found in the same rocks as *Liopleurodon,* and until proved otherwise, it is reasonable to assume that *Rhamphorhynchus* lived alongside, or rather, above *Liopleurodon.*

Dating rocks

Fossil-bearing rocks are dated in various ways. Sometimes the fossils themselves are used to date the rocks in a relative way. Some fossil species had short time-spans, but were so widely distributed and common that their remains can be used as reliable indicators of a rock's age over very wide areas. However, this dating method only works well in marine rocks, and dinosaur fossils mostly occur in non-marine rocks, usually formed in rivers and lakes.

Instead of using fossils as age indicators, actual numerical dates can be obtained using radiometric dating techniques. These techniques usually require the presence of volcanic rocks that contain minerals with radioactive elements such as potassium and uranium. The decay constants of these elements allow precise dates to be placed on the rocks, providing that the products of the radioactive decay have not escaped. Therefore, it is important that only the best-preserved of volcanic rocks be used. They must be fresh, since deeply weathered rocks will give spurious dates.

Dinosaurs in the scheme of life

Life has existed on Earth for probably four billion years, perhaps a little longer. We know this, not from fossils of tangible remains, but from geochemical signatures of life detected in the most ancient of sedimentary rocks. The first true fossils, as opposed to these geochemical signatures, occur in rocks of around three billion years old, and are traces of ancient bacteria trapped in chert, a mineral similar to flint. Metazoan life, that is, complex life in which the cells are differentiated into tissues (such as muscle and skin), did not appear until the Late Proterozoic, at probably one billion years ago. We have little evidence for the earliest of these animals and infer their existence from fossil burrows. The biological "invention" of mineralized (bone and shell) skeletons was a relatively late event, occurring simultaneously in a wide variety of organisms around 540 million years ago.

This event really marks the beginning of the "good" fossil record, as hard parts – such as bones, teeth and shells – preserve very readily. The fossil record revealed by these hard parts is the primary data set

THE TYPES OF FOSSIL

PALEONTOLOGISTS RECOGNIZE three main types of fossil: body fossils, trace fossils, and molecular fossils.

Body fossils are the tangible remains of organisms such as bones, teeth, shells, and leaves. Body fossils may be preserved as original materials, or replaced by minerals, which often make them heavier than they were in life. Such remains can be very common, and can be easily discovered where sedimentary rocks younger than about 540 million years old are exposed at the surface.

Trace fossils are marks left mainly in sedimentary rocks and result from the activities of ancient animals. The most common are walking and feeding traces, where the everyday activity of an animal can generate hundreds of trace fossils. A dinosaur will only leave one skeleton when it dies, and may leave a few hundred shed teeth during its life, but it may theoretically have generated many thousands of footprints. Other trace fossils include structures created by animals as places to live, such as burrows and nests, and also remains of fecal material, called coprolites. Coprolites are particularly important as they often preserve the remains of food within them, in the form of pieces of bone or chopped-up plant material. Unfortunately, it can be very difficult to determine which animals produced which coprolites. Dr. Karen Chin of the University of California has discovered giant coprolites in the latest Cretaceous of Montana that contain bone fragments. Since the only large predator in these deposits is *Tyrannosaurus rex*, she concludes this giant predator made the coprolites.

Molecular fossils are usually only encountered by geochemists with access to sophisticated analytical tools. They are the remains of molecules – either original molecules or breakdown products of the original molecules – that make up animals and plants. Fossil DNA is perhaps the best-known of molecular fossils, but there are many thousands of others. Although there have been claims for the discovery of fossil dinosaur DNA, no laboratory has been able to replicate them,and so at present, it is doubtful if fossil dinosaur DNA still exists.

for reconstructing the history of life on Earth. Other important data sets from which evolutionary biologists reconstruct the history and diversity of life include biological clocks and molecular fossils. Evolutionary pathways can be determined for living animals and plants by comparing their genetic make-up. Although calculating the timing of evolutionary events based on genetics is difficult, the use of genetic techniques in conjunction with fossil and geological information enables us to build up a reasonably accurate picture of the evolutionary history of life.

The remains of the first dinosaurs are found in rocks of Triassic age. That is, in rocks of between 245 and 208 million years old. Thus began the Age of the Dinosaurs. The last of the dinosaurs are found in rocks of latest Cretaceous age at around 65 million years old. Technically, dinosaurs did not become extinct because birds, the direct descendants of theropod dinosaurs, should strictly be considered a specialized group of dinosaurs. Given that there are around ten thousand species of bird alive today, it is possible to argue that we still live in the age of the dinosaurs!

▽ A *Coelophysis* eyes the skull of a *Postosuchus*. The skull will later become entombed in sediment, and then become fossilized.

The times and places in *Walking with Dinosaurs*

Dinosaur remains have been discovered on all continents, but there are places where their remains are more abundant, or exceptionally well preserved. *Walking with Dinosaurs* has concentrated on remains from these important localities. The Hell Creek and Lance formations of Late Cretaceous age in North America give us *Tyrannosaurus, Triceratops,* and the hadrosaurs. The dinosaur-bearing strata of Early Cretaceous age in Victoria, Australia, were deposited close to the Cretaceous South Pole. These rocks yield a variety of dinosaurs, thought to have inhabited cooler regions that must have had life strategies for coping with the cold and long periods of darkness. Other Early Cretaceous rocks include those of the Isle of Wight, England, where we find *Polacanthus* and *Iguanodon,* both of which appear in *Walking with Dinosaurs.* The Early Cretaceous pterosaur-bearing deposits of northeast Brazil provide *Tapejara* and *Ornithocheirus.*

For the Jurassic, the famous Oxford Clay of England is one of the best sites for marine reptiles in the world. These marine organic-rich mudrocks yield abundant remains of giants like *Liopleurodon,* the ichthyosaur *Ophthalmosaurus* and the occasional dinosaur that drifted out to sea. They also contain abundant marine invertebrates. In North America the Late Jurassic Morrison Formation is a huge deposit stretching across more than one million square miles of Colorado, New Mexico, Utah, and Wyoming. It yields the famous dinosaurs *Allosaurus, Diplodocus, Stegosaurus,* and many others.

Triassic rocks, such as those of the Chinle Formation of Arizona, have been invaluable for reconstructing life in the Late Triassic. The extensive Chinle deposits not only contain the remains of *Postosuchus* and *Placerias* but also the dinosaur *Coelophysis.* They also contain huge fossil forests of trees with trunks more than 33 feet long.

Bringing dinosaurs back to life

Advances in computer technology and animation skills developed for the film industry have allowed the animation team behind *Walking with Dinosaurs* to produce the most accurate and realistic reconstruction of life in dinosaur times ever seen. The six-part series examined the diversity and complexity of life over the vast span of the Mesozoic. While there have been many programs devoted to dinosaurs, none has placed dinosaurs into the complex and dynamic ecosystems of their time. *Walking with Dinosaurs* was the first to do this.

New fossil discoveries mean that paleontologists are continually reassessing their interpretations of the fossil record. The *Walking with Dinosaurs* team wanted the series to be as accurate and up-to-date as possible. Although they drew heavily from what is already known about the appearances and lifestyles of fossil animals, the team also introduced new information as yet unpublished. This "on the ball" approach meant that the programs were as scientifically accurate as possible. Even so, during the making of the series, new discoveries were made that came too late to be included. For example, at least four new species of pterosaur, the flying reptiles of the Mesozoic, were discovered while *Walking with Dinosaurs* was being put together. One of these discoveries is larger even than *Quetzalcoatlus*, the giant of the skies featured in *Death of a Dynasty*.

Nevertheless, for many dinosaurs and other Mesozoic animals we know their age, their diet, their prey, their predators, their environment, and for some, we know about their sex life. We have growth histories from egg to adult, and we can calculate their walking and running speeds from their footprints. From these same footprints we can also determine the style of their gait. This is of paramount importance for animators, allowing them to recreate the walking and running styles of the various dinosaurs. Combining this factual information with inferences from the living world, a realistic picture of life in Mesozoic times was created for the series.

Reconstructing the life appearances of fossil animals

One of the first steps in bringing organisms back to life for their performances in *Walking with Dinosaurs* was to reconstruct the life appearances of the fossil animals and plants. For dinosaurs the basic information regarding size and shape comes from the fossilized skeleton. For extremely well-preserved skeletons, where all of the bones are preserved in their original positions, the task is relatively simple. For skeletons of scattered bones the task is more difficult, and for incomplete skeletons it is more difficult still. Piecing back together

▷ **In reconstructing the pterosaur *Ornithocheirus*, experts rely on information from several closely related pterosaur types.**

a dinosaur skeleton is like doing a giant jigsaw, but because of the distortion that is frequently found in fossil bones, the pieces do not always fit perfectly and many crucial pieces can be missing – and of course the bones do not come in a box with a picture on the lid!

Once the skeleton has been reconstructed, the degree of movement of all of the bones is determined from the size and structure of the joints between them, and an assessment of the limitations imposed on these by ligaments and muscles. The outline of the animal with its soft tissues is ascertained by reconstructing the muscle mass as determined by what we know about the anatomy of living animals. Muscles can be identified by the scars they leave on the surfaces of bones at their points of attachment, so their location and size can be determined, but this is not always easy.

Once the muscle mass has been reconstructed it is a relatively simple task to put a skin over the surface. However, fossil evidence shows that the skin of dinosaurs varied in texture between dinosaur

▽ **The serrated frills seen here on the back of *Diplodocus* were unknown until recently and are a rare example of exceptional preservation.**

types, and in many groups the skin contained rows of bony scutes (scalelike structures), spines, and plates, as in the famous *Stegosaurus.* There is and has always been controversy over the distribution of the plates and spines of stegosaurs, and much attention must be paid to the excavation of these dinosaurs so that the exact position of the plates and spines can be mapped out.

Reconstructing dinosaur behavior

Walking with Dinosaurs shows dinosaurs behaving in ways comparable with behavior found in living animals. By the very nature of the fossil record, we will never have as much information about the lifestyles of ancient animals as we do about living ones. Such things as color and sound are absent from the record, as are most key details of the way in which animals behaved. Consider the complexities of modern animal behavior – mother crocodiles carry their babies to the water; they protect them from predators; and they respond to their distress calls. How can we know that mother dinosaurs behaved in similar ways? We can't. However, in a few special cases, behavior has been preserved in fossils. For example, some small dinosaurs are preserved sitting on top of their nests and others are preserved locked in combat with other dinosaur species. Dinosaurs such as the Jurassic *Compsognathus* have been found with food still in their stomachs, providing clear evidence of their feeding activity. In this case, *Compsognathus* ate small lizards as part of its diet.

But for the most part, reconstructing behavior in fossil animals is inference and informed speculation, inspired by the behavior of the animals of today. At its simplest level, this kind of behavioral reconstruction is demonstrated in *Walking with Dinosaurs* by the portrayal of the giant alligator *Deinosuchus* as an amphibious predator that ambushes land animals from the water's edge, a scene portrayed in *Death of a Dynasty.* This is an analogy drawn with the behavior of large modern crocodiles and alligators, which *Deinosuchus* resembles.

When modern counterparts don't exist, as with the long-necked sauropod dinosaurs, behavior becomes more difficult to reconstruct. Historically, early dinosaur workers thought that sauropods were too large to walk on land, and thus many early artistic reconstructions depict them wading in lakes, their huge mass supported by the water and only their heads and necks projecting above the surface. We now know that the weight of the water would have prevented them from breathing, and biomechanical studies show that they were perfectly capable of walking on land. These days, paleontologists are more concerned with things like whether sauropods could or could not elevate their necks like giraffes or stand on their hind legs as elephants occasionally do.

How do we know what colors extinct animals were?

The animals of *Walking with Dinosaurs* display a variety of color schemes, ranging from the dull, patternless gray and brown of the *Diplodocus* to the bright blue and red of the small bird *Iberomesornis.* Bright color patterns were also used for some of the pterosaurs, and bold white stripes and a striking black, white and gray pattern was used for the small Cretaceous mammal *Didelphodon.*

Again, if we were to stick solely to solid evidence, we would have to reconstruct nearly all fossil animals as being completely colorless and devoid of any skin patterns. In a few exceptional cases, fossil colors are preserved, such as on the iridescent wing cases of some fossil beetles and on fossil scallop shells, and color patterns (but not original colors) are known for some fossil feathers.

By far our greatest guide to the colors and patterns of fossil animals comes from the great range and diversity of colors and patterns seen in living animals. Living animals show us that there are a few fairly consistent rules in animal coloration. Giant animals tend to be dull-colored, making it likely that giant dinosaurs were too. Most aquatic animals exhibit what is known as dorso-ventral counter-shading where the underside is light-colored and the upper side is dark-colored. This counter-shading camouflages the animals when they are seen either from above (and are therefore seen against the dark background of the seafloor), or from below (when they are seen against the light background of the water surface). We can therefore be fairly confident that aquatic animals of the past, including fish, the fish-shaped reptile *Ophthalmosaurus,* and the giant predator *Liopleurodon,* were counter-shaded in similar ways. Many large aquatic animals, including many of the sharks, bony fish, dolphins, and whales, also have markings such as spots, stripes, and dark lines on their flanks. These may act as recognition markers for other members of their species, or they may help to break up their outline and so help camouflage them from predators. Again, therefore, it was a reasonable deduction to give the aquatic animals in *Walking with Dinosaurs* similar color schemes.

Similarly, the use of color patterns that camouflage land animals today might have worked in the same way in the past, either to allow the animal to blend into a particular background by being a similar color, or to disappear by breaking up its outline. Thus *Iguanodon* in *Walking with Dinosaurs* was given stripes that are similar to those of the living forest-dwelling okapi.

How do we know what sounds extinct animals made?

The animals in *Walking with Dinosaurs* make varied noises. From the smallest pterosaur to the enormous *Diplodocus* and *Liopleurodon*, they grunt, roar, purr, hiss, bark, and rumble. How do we know that they made these noises? The answer, quite simply, is that we don't and

STRIKING COLORS IN DINOSAUR TIMES

IT IS COMMON for display structures in animals to be brightly colored. In this way they make the animal more attractive to the opposite sex and convey information about its health, breeding status, and maturity. Examples include the bright red wattles and crests on cockerels, the skin frills and head flaps of some lizards, and the brightly colored muzzles and backsides of baboons and mandrills, among many others. Certain structures in fossil animals appear to have been comparable with such display features and may reasonably be regarded as having been for sexual display. However, this can only be determined if both sexes of a fossil species are found and are clearly somehow different in an aspect of their physical make-up. Recognizing sexual variation among fossils is difficult because, if the two sexes are very different, it is likely that paleontologists will regard them as two different species!

Many of the animals featured in the documentary *Walking with Dinosaurs* possessed structures that appear to have evolved for the purposes of sexual display. It is therefore reasonable to make these structures brightly colored. As an example, the nose horn of the small Jurassic dinosaur *Ornitholestes* was given a bright blue color.

△ The striking red horns of *Allosaurus* are modeled on features seen in living reptiles and birds.

probably never will. However, noise is made by most kinds of living animal and it is certain that animals known only as fossils also made noises when they were alive. Animals make noises to attract mates, to distract or repel predators, and to communicate for various purposes with other members of their own species. That ancient animals are known to have had ears is a clear indication to us that animals were making noises in the Mesozoic.

Can fossil sound be preserved? The answer must be no, but a handful of attempts have been made to scientifically reconstruct the noises that fossil animals would have made. For example, it is technically possible to recreate the noises made by some fossil insects. The organs that these insects use to make sounds are often preserved, therefore a replica could be used to reproduce the noise, or at least something similar. However, we would not know the speed at which the noise was made, and hence its frequency, or the pattern of the

▽ **Reconstructing the behaviors and lifestyles of extinct animals, like this predatory allosaur, is a difficult and controversial area that relies on much inference and speculation.**

sound. We would not know if the song of the animal was one long sound or several short bursts.

With regard to the noises made by dinosaurs, much work has been done in reconstructing the calls that might have been made by hadrosaurs, a group of herbivorous dinosaurs from the Late Cretaceous. Some of these animals had hollow crests on their skulls. It has been assumed that these crests must have had a function, and that one of these functions may have been in amplifying or resonating sound. Of course, they may have been multifunctional, and sound production or modification may not have been their primary purpose. Nevertheless, Dr. David Weishampel of John Hopkins University, Baltimore, has recreated sounds from models of the crests of hollow-crested hadrosaurs. Apparently the hollow crests produced low, trombonelike noises.

Many animals produce vocalizations without the use of recognizable hard parts. Birds sing using tubes in their lungs as well as a muscular structure in the throat called the syrinx. Both of these have little or no preservation potential, and, therefore, even in the best-preserved fossils, we are unable to state whether or not Mesozoic birds could sing. But animals don't always need modified anatomical structures to generate sound. Woodpeckers, for example, produce percussion sounds by pecking on tree trunks, and whales hiss while expelling water from their nostrils. Thus, even if we do not find any skeletal evidence for sound production, we should not rule it out for any particular fossil animal.

The sounds in *Walking with Dinosaurs* were chosen, not at random, but with due regard for the size of the animals and the environment in which they lived. The giant sauropods, like *Diplodocus*, were reconstructed as using low frequency rumbling calls based on those of elephants, while the small pterosaurs were given calls inspired by those of modern seabirds.

Some dilemmas

There are many things that we do not know about Mesozoic life, some of which are discussed above, such as color and sound. Paleontologists by and large confine themselves to studying facts and trying to interpret fossils in a thoroughly scientific way. Only rarely do they speculate on the life habits of fossils. Thus, most scientific papers on dinosaurs are very dull affairs describing in great detail the bones of the skeleton and the manner in which they functioned. Some are analyses of mechanical properties of bones derived from mathematical formulas. Only occasionally are remains discovered which allow insight into the life habits of dinosaurs. The team of scientific advisors on *Walking with Dinosaurs* were entering a new world in which speculation on life habits was essential.

At times it is quite possible that the animators thought the scientists must have been pretty brainless. On many occasions the scientists were unable to answer the simplest of their questions. How fast could *Liopleurodon* swim? Fortunately for this one there were some calculations available, although the error bars were quite large: it seemed that *Liopleurodon* was generally a slow swimmer, but capable of short bursts of speed. *Liopleurodon,* unlike dinosaurs, cannot leave footprints, and so there is nothing tangible to determine its swimming speed, except by analogy with modern animals. In the end it was the animators who had to give it a speed that simply looked right to them.

For many of the animals portrayed the data sets are very poor. There is for example, no complete skeleton of the giant newtlike *Koolasuchus,* but as a large predator, it was an important and unusual animal in the southern Cretaceous ecosystem in which it lived. Not to have included *Koolasuchus* would have been to miss out a vital component of the high latitude Cretaceous polar world.

Because the series was made as a "fly on the wall" documentary there could be no comments in the narration such as "*Tyrannosaurus may* have done this," or "*Diplodocus may* have done that." Such commentary would have been at odds with what was shown as actually happening. In any documentary on living animals the animals themselves write the story line. For a documentary on the migration of reindeer, for example, it is the deer that decide where they are going to go. For *Walking with Dinosaurs,* fossils could only write part of the story line: the producer had to fill in the gaps.

Walking with Dinosaurs brings to life some of the most amazing animals that have ever lived and portrays them as vital, living, breathing animals—not, as many people think of them, as dusty bones in dingy museums. Let's hope that this new genre of prehistoric "documentary" continues to flourish: long may dinosaurs walk.

CHAPTER 1

New Blood

220 MILLION YEARS AGO

THE DINOSAUR STORY starts in the Triassic period, a time in which the world was radically different from that of today. All continents were united into the huge supercontinent Pangaea (meaning "whole earth"), a landmass whose size and relative lack of coastline would have had profound effects on its climate. Large areas of Pangaea were arid or semi-arid and rainfall was erratic and perhaps seasonal. This environment clearly favored the evolution of reptiles, animals able to conserve water by excreting their waste products in the form of dehydrated pastes and crystals. However, most fossil assemblages of the Triassic come from wetter areas where water and vegetation were abundant, because there dead plant and animal remains stood a better chance of being buried by sediment, and hence, of being fossilized.

A select handful of specific locations scattered around the world allow us to reconstruct Triassic life. One location that has proved of tremendous significance is the Chinle Formation in the southwestern USA. It is this area and its animals that form the focus for the Triassic episode, *New Blood*, of *Walking with Dinosaurs.*

Animals of the Late Triassic – creatures from a time of change

Various Chinle animals were reconstructed in detail for the *New Blood* episode. These are important since the Late Triassic was a time of great change in the composition of animal communities. Dinosaurs and the ancestors of mammals, the cynodonts, were to become important, and of course the rise of the dinosaurs sets the stage for the rest of the whole Mesozoic Era. Members of these groups therefore play a major role in the *New Blood* episode. *Coelophysis*, a nimble, two-legged carnivore, represents the early dinosaurs, while *Thrinaxodon*, a cat-sized animal that hid in burrows, plays the part of the cynodonts. In this chapter we examine the evidence that resulted in the reconstructions and in some of the behaviors depicted in *Walking with Dinosaurs* for these animals.

◁ The abundance of specimens at some key localities indicates that dicynodonts like *Placerias* were herding animals. Perhaps in this way they were protected from predators like *Postosuchus*.

Besides early dinosaurs and cynodonts, we also look here at the dicynodonts (represented in *Walking with Dinosaurs* by the tusked *Placerias*) and the giant predator *Postosuchus*. These and many other kinds of animals were destined to become extinct and are therefore depicted in *Walking with Dinosaurs* as the last remnants of once flourishing groups. The first of the flying reptiles – the pterosaurs – appeared in the Late Triassic and are also looked at in detail here. At the end of the Triassic, truly giant dinosaurs had appeared and heralded the future of the Age of Dinosaurs. The dinosaur that was reconstructed for the *New Blood* episode was the herbivore *Plateosaurus. Plateosaurus* is one of the best-known Triassic dinosaurs and thus much information was available in reconstructing it. With all of these animals and others, the Late Triassic was a truly dynamic time, and one of the most important in the history of life.

Competitors for early dinosaurs – the crocodile-group archosaurs

Dinosaurs were present worldwide by the start of the Late Triassic. However, they were not very diverse and not particularly numerous. The group of reptiles to which dinosaurs belong, the archosaurs ("ruling reptiles"), were represented at this time by a fantastic array of forms besides dinosaurs. Rather than being close relatives of dinosaurs, many of these belonged to the same group of archosaurs as the living crocodiles. These crocodile-group archosaurs have strikingly different anklebones from dinosaurs and tend to be long and low animals without the capacity for sustained bipedal (two-legged) movement. During the Middle and Late Triassic, crocodile-group archosaurs dominated freshwater environments in the shape of the phytosaurs, superficially crocodile-like archosaurs with long skulls and nostrils in front of their eyes. True crocodiles had evolved, but rather than being the large, amphibious predators we know today, they were small, nimble, terrestrial reptiles that probably hunted insects and other small animals. Also on land, aetosaurs were heavily armored herbivorous crocodile-group archosaurs that sported spikes, horns and armor plates along the top and sides of their bodies.

A diverse array of terrestrial carnivorous crocodile-group archosaurs with deep skulls and serrated, recurved teeth were the ecological precursors of the predatory dinosaurs, the theropods – the group that would be the dominant predators later on. Unlike

theropods, these Triassic archosaurs were quadrupedal (that is, walked on four legs) with forelimbs and hindlimbs of nearly equal length and hands suited for weight bearing.

One such predator, and probably the dominant one in the place in which it lived, was *Postosuchus*, a huge archosaur from Texas, Arizona, and New Mexico and named after the small town of Post, in Texas. In *Walking with Dinosaurs*, *Postosuchus* was characterised as a huge and formidable predator that looked something like a cross between a crocodile and a *Tyrannosaurus*. Reaching 20 feet in length and with a huge, deep skull, *Postosuchus* appears to have been an awesome predator that could tackle and dispatch, with relative ease, most of the large animals of its time.

◁ **The head of a killer. *Postosuchus* had two ridges running along the top of its head. Its skull was deep and narrow and it had sharp, recurved teeth.**

What did *Postosuchus* look like, and how did it move?

Well-preserved *Postosuchus* fossils allowed this animal to be reconstructed for its role in *Walking with Dinosaurs*. The skull of *Postosuchus* is notably narrow from side to side, with elongated ridges running along the top of each side of the skull. Perhaps these were adorned with hornlets and used as display structures in courtship and aggression. Like other crocodile-group archosaurs, the neck, back, and tail of *Postosuchus* were covered in armored scalelike structures called scutes. It is often difficult to figure out the exact arrangement of scutes in extinct reptiles because they are so rarely preserved in place. We can be confident that *Postosuchus* and the archosaurs like it were covered in pebblelike rounded scales, not only because their closest living relatives, the crocodiles, have such scales, but also because some fossil trackways made by *Postosuchus*-like animals preserve skin impressions. Such footprints from the Triassic of Italy, described in 1999 by Dr. Marco Avanzini, show that these archosaurs had polygonal scales covering the palms of their hands and soles of their feet.

In *Walking with Dinosaurs*, *Postosuchus* is shown walking on all fours but rearing up on it back legs when engaged in an aggressive encounter. This reconstruction reflects years of controversy that have surrounded the posture and locomotion of *Postosuchus*. Professor Sankar Chatterjee from Texas Tech University, the paleontologist who named and first described *Postosuchus* in 1985, argued that it had proportionally small, slim forelimbs and was predominantly bipedal. In effect, Chatterjee regarded *Postosuchus* as superficially similar to later

◁ **The biggest carnivore of its day, *Postosuchus* was a 20-foot predator that walked on all fours and had tall legs. A member of the "crocodile group" of archosaurs, *Postosuchus* had bony armor covering its back and tail.**

giant theropods like *Tyrannosaurus*. A more recent analysis by Drs. Robert Long and Peter Murry suggests an alternative. Chatterjee's reconstruction of an adult *Postosuchus* included scaled-up material from smaller animals of between six and ten feet in length, interpreted by Chatterjee as juvenile *Postosuchus*. However, Long and Murry found that these "juveniles" were actually the remains of a different, but related, animal. This differed from *Postosuchus* in having an elongate, slender neck, fundamentally different hips with a different number of hip vertebrae from *Postosuchus*, and longer, slimmer limbs. Long and Murry named this new animal *Chatterjeea* in honor of Chatterjee's work.

Forelimb material from the large, "true" *Postosuchus* individuals reveals a well-muscled upper arm but relatively small hands. Though estimated to be slightly shorter than the hindlimbs, these forelimbs are still long enough, and robust enough, for *Postosuchus* to have used them for walking on. Similarly proportioned forelimbs are seen in related archosaurs that have never been regarded as bipeds. Also, fossil tracks from the Early and Middle Triassic, named *Synaptichnium*, appear to have been made by an animal related to *Postosuchus*, and reveal hand as well as foot impressions. However, it still appears possible that *Postosuchus* might have run bipedally on occasion, and perhaps it reared up to intimidate enemies or when fighting or displaying in courtship.

Direct evidence that archosaurs like *Postosuchus* fed on other large Triassic animals comes from bite marks preserved on their bones. *Postosuchus* was not built for speed, but had longer, slimmer legs than the large, bulky herbivores of the time and may have hunted them by rushing out from cover and inflicting fatal wounds on the side of the body. It would clearly have been an awesome predator that smaller animals, including dinosaurs of the time, would have tried to avoid.

Coelophysis, an early predatory dinosaur

Early in the Late Triassic, dinosaurs were small and rare. Though the very earliest known dinosaurs were not featured in *Walking with Dinosaurs*, *Coelophysis* played the part of an early dinosaur living in a world ruled, at first, by other kinds of animal. Though dinosaurs were rare at this time, new discoveries in the Late Triassic rocks of North and South America, Madagascar and elsewhere show that they had diversified into their three major groups. By the end of the Late Triassic, clearly differentiated forms of all three groups were around and had adopted fairly different lifestyles. *Coelophysis* was one of the theropods, the bipedal, mostly predatory group of dinosaurs that have curved claws on their hands and functionally three-toed feet.

Juvenile specimens of *Coelophysis*, including half-grown specimens and even hatchlings, are known. They show that, as is true of most backboned animals, juveniles had larger heads, proportionally bigger eye sockets and shorter legs than adults, but that these proportions changed with growth. That these juvenile skeletons are preserved together with adults might suggest that there was active parental care in these dinosaurs.

Was *Coelophysis* a cannibal?

In stark contrast to this image, it is often thought that *Coelophysis* was a cannibal, at least on occasion. This behavior was reconstructed for *coelophysis* in *Walking with Dinosaurs*. How confident could scientists possibly be in reconstructing cannibalism in long-extinct dinosaurs? Surprisingly, two adult specimens, preserved in the collection of the American Museum of Natural History in New York, appear to preserve juvenile *Coelophysis* specimens in their stomach regions. Cannibalism of this kind would certainly not be unexpected among carnivorous dinosaurs – in living animals, virtually everything from fish, snakes, and lizards to big cats, eagles, and crocodiles practices cannibalism – and so, we would therefore predict its occurrence in dinosaurs such as *Coelophysis*.

However, some doubt now surrounds the idea that the two New York *Coelophysis* specimens really do contain juveniles within their skeletons. Further examination suggests that the adult specimens are simply lying on top of the juvenile skeletons. Detailed investigation is needed to determine whether or not this is correct. If it is, the idea of cannibalism in *Coelophysis* lacks material evidence at this time.

With its elongated, slim skull and sharply serrated, recurved teeth, plus its relatively small body size, *Coelophysis* might have specialized in capturing lizardlike reptiles and perhaps fish and amphibians. At the front of its lower jaw, its teeth radiated outwards somewhat while those at the front of the upper jaw were separated from the teeth further back by a notch in the upper jawline. These specialized jaws appear

HOW TO DETERMINE A DINOSAUR'S SEX

COELOPHYSIS IS FOUND as two differently shaped types. One is larger, with a proportionally longer head and neck but shorter arms and tail; the other is smaller, with a proportionally smaller head and neck and longer arms and tail. It seems obvious that these two forms reflect sexual variation, but which form is the male, and which the female?

In most mammals, lizards, and crocodiles, males are usually larger than females. Maybe this was the case for *Coelophysis*. In turtles and some living birds, however, particularly birds of prey, it is the females that are larger. Some scientists have suggested that birds of prey might make good comparisons with predatory dinosaurs, and that it would be logical if the larger *Coelophysis* specimens were actually the female ones. Though much speculation has surrounded this issue, it was until recently untestable: that is, there was no way of proving which of the two alternatives is really the correct one.

A new technique, developed by the German paleontologist and zoologist Dr. Dino Frey, might finally resolve the matter . Frey found that male crocodiles have a larger first tail chevron than females. Chevrons are rod-shaped bones that point downwards from the bottom of the tail vertebrae.

First chevrons that are larger in males than in females might be explained by two functions. First, the male might need his larger first chevron as an anchor for his penis muscles. Second, the female would benefit from a short first chevron as it allows more space for an egg to pass. Among dinosaurs, these differences in chevron size have been examined for the Late Cretaceous theropod *Tyrannosaurus*. It appears that the specimens with the shorter first chevrons are actually the bigger, more robustly built specimens, so perhaps theropods were like turtles and birds of prey after all. Assuming that the two forms of *Coelophysis* do represent the sexes, perhaps the females were stronger, more aggressive and more dominant than the males.

△ The lightweight, agile predatory dinosaur *Coelophysis* – a slim-headed hunter known from hundreds of specimens found in the Late Triassic sandstones of North America.

◁ **Because cannibalism is widespread in modern predators, it seems reasonable to think that predatory dinosaurs would have practiced it. Here an adult *Coelophysis* eats a juvenile.**

to have been ideal structures for grabbing small animals. Direct evidence for *Coelophysis'* diet, however, is lacking. It is also possible that *Coelophysis* was a big-game hunter that could attack and weaken large animals, like dicynodonts and herbivorous dinosaurs, by taking repeated slashing bites from their legs or sides.

Dinosaur success in the Triassic

In *Walking with Dinosaurs* we see how dinosaurs were at first rare, then eventually became abundant and more diverse. Though it is clear from *Walking with Dinosaurs* that early dinosaurs were agile predators, was their eventual success guaranteed, or was it thanks to fortuitous changes in the environments of the time, as the series depicted? This area has been the subject of argument for many decades. Following studies of the proportions of different kinds of animal in Triassic fossil assemblages worldwide, most scientists agree with the model proposed by Professor Michael Benton in 1983. Rather than believing that dinosaurs had actively competed with other kinds of Triassic animal, Benton proposed that dinosaurs were "victors by default."

Previously, some scientists had argued that dinosaurs had been ecologically superior to the other Triassic animals, being better equipped to hunt and kill. Problems with this idea include the fact that some non-dinosaur reptiles also had the same special features as dinosaurs, yet did not later become successful. Also, though dinosaurs first evolved in the middle of the Triassic, they did not become dominant until a mass extinction at the end of the Triassic destroyed the competition. Surely if dinosaurs were "superior" would they not have risen to dominance much sooner? Dinosaur success was not ensured by the special features that set them apart from other archosaurs, nor were Triassic dinosaurs particularly better at killing or eating than other animals of the time.

Placerias and its relatives – the dicynodonts

While archosaurs like *Postosuchus* were the dominant predators at this time, the role of terrestrial herbivore was still largely controlled by the dicynodonts. These were large piglike beasts with horny beaks and two large, protruding tusks at the front of the upper jaw (dicynodont means "two dog-teeth"). *Placerias*, a bizarre, tusked, herbivorous animal that appears to have lived in herds, represents the dicynodonts in the *New Blood* episode of *Walking with Dinosaurs.*

Dicynodonts were quite unlike any animals alive today, but one expert has suggested that they might be imagined as "turtle cows". Their hindlimbs were stout and erect while their massively muscled forelimbs splayed out sideways. Dicynodonts were not reptiles but part of a major assemblage of animals called synapsids. Earlier synapsids, from the Permian age, include the sail-backed *Dimetrodon*. Later representatives include mammals; thus we humans are also synapsids.

It is clear from their abundance that, in the first part of the Triassic, dicynodonts were phenomenally successful. *Lystrosaurus*, a small dicynodont from the earliest Triassic of South Africa, India, China, Russia, Australia, and even Antarctica, is often the most common fossil in the assemblages in which it occurs and hundreds of individuals are known. The discovery of *Lystrosaurus* – a small, apparently slow-moving animal, probably without the ability to make long sea crossings – in these multiple far-flung locations was one of the key pieces of evidence originally used to support the theory of continental drift. When *Lystrosaurus* was alive, these areas were all connected, thus explaining its distribution.

How do we know what dicynodonts ate?

Placerias in *Walking with Dinosaurs* is reconstructed as a low-browsing herbivore that eats plants and digs with its tusks and beak for roots, not unlike some members of the pig family. Various lines of evidence indicate on what, and how, *Placerias* and its relatives fed. The beaks and tusks of dicynodonts, and the bulky shape of their bodies, indicate that they were herbivores that grabbed and bit at leaves and plant stems. Their teeth and beaks preserve wear marks that seem to have resulted from contact with abrasive plants, and some forms are thought to have used their enlarged tusks to dig for roots and tubers.

Like a variety of other herbivorous animals, dicynodonts seem to have had the ability to move their lower jaws backwards and forwards. Once plant food was bitten off by the beak, they would then have used this jaw motion to shear it into smaller pieces. The beaks of some dicynodonts, however, are not unlike those of some living turtles in shape. Seeing as such turtles may be omnivorous or even carnivorous, paleontologists Dr. Hans-Dieter Sues and Dr. Robert Reisz have recently pointed out that there is no reason why at least some dicynodonts could not have been omnivores or carnivores too. The extreme abundance of forms like *Lystrosaurus* and *Placerias* does imply, however, that they were herbivores: carnivores cannot usually thrive at such high densities because they cannot find enough food.

Some fossils of Early Triassic dicynodonts are actually preserved inside vertically spiraling burrows, indicating that they had evolved behaviors that helped them to shelter from the extremes of the Triassic climate. Such burrowing forms were small, however, being less than 20 inches long; the possibility of burrow dwelling does not seem to have been open to the larger, later dicynodonts like *Placerias*. These were giants with skulls about two feet long and with a body about ten feet long and a ton in weight.

The last dicynodonts

Despite their earlier success, dicynodonts were rare by the end of the Triassic and only represented by a handful of species. *Placerias* appears to be the last one and is consequently referred to in *Walking with Dinosaurs* as an "endangered species." Some experts have suggested that some of these last species were ecological recluses, something like the living giant panda, which only survived by inhabiting remote areas where few other large creatures could thrive, and

▽ Like all dicynodonts, *Placerias* was a heavily built creature with stout limbs and a rounded body. Its beak and prominent tusks would have made short work of most vegetation.

by living a secretive, specialized lifestyle. This does not seem to have been true of *Placerias*, as it is preserved in abundant groups in what were once wet, well-vegetated areas.

How did *Placerias* live and behave?

Exactly how *Placerias* and the other last dicynodonts behaved is still somewhat mysterious. Their barrel-chested shape and spreading feet have suggested to some that they may have been adept in water and partially amphibious, like hippos. Thus it is reasonable to think that they wallowed in lakes and rivers, as is shown in *Walking with Dinosaurs*. The moist environment in which they seem to have died also supports this amphibious idea. It is also possible that the animals are preserved in this environment because they were seeking water during a time of drought, and then either died there from dehydration or from becoming mired in mud, or fell prey to predators.

Virtually all of the world's known *Placerias* specimens come from a quarry in the Chinle Formation of Arizona. These specimens are not articulated: their bones are scattered with many of the elements missing, but, strangely, the large skulls are represented in abundance. Usually it is the other way round in fossil assemblages – skulls are very rare or totally absent, even when skeletons are well represented.

◁ Wear marks on their tusks show that *Placerias* and their relatives sometimes dug in search of plant roots. Fossils of *Placerias* are abundant at one Arizona locality, thought to represent a dried up lake bed.

Dicynodont expert Dr. Gillian King suggested that this style of preservation implies scavenging by predators which preferred eating body parts to the hard, bony *Placerias* skull.

The abundance of *Placerias* skulls has allowed paleontologists to analyze the variation within the population, something rarely possible for fossil animals because the number of known individuals is nearly always very low. Two forms seem to be present, distinguished by the size and shape of the downward-facing pointed structures that house their tusks. It is assumed that the males were the ones with the bigger structures, and that they may have used these in shoving matches when competing with one another.

Cynodonts – "proto-mammals" from the early days of the dinosaurs

One of the main characters in *New Blood* was a species of burrow-dwelling cynodont, a vaguely doglike animal that seemingly combined traits of reptiles and mammals. Along with many other small Triassic animals, such creatures lived in the shadow of the Triassic archosaurs and the large dicynodonts. From our point of view, however, cynodonts are extremely important because they include the ancestors of mammals. In fact, according to the technical definitions preferred by some experts, cynodonts *include* mammals, and thus we are cynodonts as much as the small "proto-mammals" of the Triassic. Though the Triassic cynodonts were not true mammals, and had not yet evolved the specialized lower jaw, ear bones, and other features seen in true mammals, it seems probable that at least some of their behavior would be recognizably mammalian.

Cynodonts first appeared at the end of the Permian and diverse varieties evolved in the Triassic, several of which were herbivores. One of the best-known Triassic forms is *Thrinaxodon* from South Africa and Antarctica, a carnivore roughly the size of a house cat and with prominent canine teeth. Because of their importance in the evolution of the first mammals, the anatomy of cynodonts like *Thrinaxodon* has been well studied and much is known about the structure of their ears, jaws, and teeth – structures that later proved key to the spread and success of mammals.

CYNODONTS: WARM OR COLD BLOODED?

ONE AREA OF CYNODONT biology that has often been controversial is whether the animals were able to generate their own body heat - that is they were "warm blooded" like living mammals - or whether they had to rely on external sources of heat, like living lizards. A possible key to this debate is provided by scroll-like bony structures found inside the snouts of mammals and birds, called the respiratory turbinates. "Warm blooded" animals have a higher rate of respiration than other animals and therefore breathe faster during and following exercise. This heavier breathing means that there is a danger of moisture loss in the exhaled air. "Warm blooded" animals seem to have overcome this problem by growing delicate turbinate bones which trap and retain moisture so that it is not lost from the body.

Analysis of the skulls of a variety of early synapsids, including cynodonts, by Dr. Willem Hillenius shows that these animals did possess respiratory turbinates, and they therefore appear to have been "warm blooded." Not only would this have given them improved stamina when foraging and hunting, it may have allowed them to be active throughout the night when it was cooler. Of course, if early cynodonts were "warm blooded" it is tempting to think that they might have been furry so that they could retain some of their body heat. Fossilised fur would be needed to confirm this, but alas, no early cynodont fur has yet been discovered. Small pits on the snouts of fossil cynodonts were once used as evidence for the presence of whiskers, and this would in turn suggest the presence of fur. However, these pits have since been shown to be holes for blood vessels.

How did cynodonts live?

Despite these many technical studies on the form and function of *Thrinaxodon*'s bones, little information on its behavior and lifestyle is available. We do know that small cynodonts like *Thrinaxodon* made their dens in burrows and that they apparently rested and slept in a curled up posture, as do modern rodents and cats. An amazing *Thrinaxodon* skeleton from South Africa, described in 1958, is preserved in a perfectly curled up position, with the head nestled against the ribcage and the end of the tail curled over the top of the head. Presumably this individual died while sleeping in its burrow.

Next to nothing is known about reproduction and parental care in early cynodonts. Seeing as the living platypus and echidna, which are true mammals, still lay eggs, it appears likely that the ancestors of mammals were also still egg layers. It is likely that cynodont parents incubated these eggs by curling around them, and that they practiced parental care. Again, an exceptional *Thrinaxodon* specimen sheds some possible light on this – it is a nodule containing an adult and a tiny juvenile specimen preserved side by side. Perhaps these represent a mother and a young baby.

Thrinaxodon in *Walking with Dinosaurs* is animated with a stiff back and slight waddle to its walk. This reconstruction stemmed from

▽ **Fossil cynodonts have been found where adults are preserved lying adjacent to small juveniles, suggesting parental care in these animals. This strategy proved invaluable to the small mammals that descended from such cynodonts.**

research on the probable appearance of this animal. As reconstructed by artists, cynodonts like *Thrinaxodon* often look like small dogs, but with more sprawling limbs. Dr. Diane and Dr. Kenneth Kermack, two experts on the evolution of cynodonts and early mammals, pointed out that such similarities are rather misleading. Living mammals, including dogs and other carnivores, have quite flexible bodies and can bend and twist when running, jumping, or climbing. Early cynodonts, in contrast, had rather stiff spines and flattened, overlapping ribs. These features would have restricted their movement: their bodies would have been stiff and they would not have been capable of much side to side movement.

▷ **It seems likely that, like primitive mammals today, cynodonts dug and rested in burrows. Some are preserved in a curled-up position, implying that they died while sleeping underground.**

Flying reptiles of the Triassic

While dinosaurs ran and walked on the ground and cynodonts rested in burrows beneath it, other forms of Triassic life were taking to the air. Even before the Triassic, earlier forms of reptile had taken to the air by gliding on skin suspended between elongated, foldable ribs or riblike structures. However, the first reptiles to evolve true flapping flight appeared in the second half of the Triassic.

Pterosaurs, winged reptiles best known for their giant Cretaceous representative *Pteranodon,* certainly had evolved by Triassic times and are now known from Triassic rocks of Europe and North America.

▽ One of the earlier pterosaurs, *Peteinosaurus* was a small long-tailed pterosaur known from Europe. Its multiple small teeth suggest that it was an insectivore.

One early pterosaur in particular, *Peteinosaurus*, was focused on for the *New Blood* episode of *Walking with Dinosaurs*.

The appearance of *Peteinosaurus* and other pterosaurs

Reconstructing *Peteinosaurus* and the other pterosaurs for *Walking with Dinosaurs* was problematic – how pterosaurs looked when alive is a highly controversial area.

A popular idea in the 1980s was that pterosaurs had very slim wing membranes that extended from the end of the wing finger to the hip, but were not attached to the leg. This reconstruction, if correct, might have meant that pterosaurs could sprint bipedally on the ground, much like small dinosaurs. Recent interpretation of exceptionally well-preserved pterosaur specimens (still with their wing membranes preserved) show that an older model, where the wing membranes are more extensive and involve most of the hind limb as they do in bats, is more likely to be correct. Work on the small Jurassic Russian pterosaur *Sordes* by Dr. David Unwin and Dr. Natasha Bakhurina reveals that these pterosaurs did have wing membranes that reached their ankles, and had a membrane in between their hind legs as well, again as there is in most bats. A similar configuration appears to have existed in the Jurassic German pterosaurs *Pterodactylus* and *Rhamphorhynchus*.

WHAT ARE PTEROSAURS?

EXACTLY WHICH GROUP of reptiles pterosaurs evolved from remains controversial. One school of thought is that pterosaurs are very close relatives of dinosaurs, and that both groups evolved from a small, bipedal archosaur adapted for fast running on the ground. Another theory is that pterosaurs evolved from small, early quadrupedal (four-legged) archosaurs that took to climbing trees and jumping from branch to branch, or from the tree to the ground. This is called the "arboreal leaping" theory and is similar to the one proposed to explain the evolution of flight in the first bats.

Yet a third idea is that pterosaurs belong to a group of reptiles called protorosaurs. These are not archosaurs, but related to them. Though most protorosaurs are superficially lizardlike and appear best suited for foraging quadrupedally on the ground, at least some could run bipedally and some appear well suited for climbing trees. Unfortunately, the very earliest known pterosaurs are still not primitive enough to reveal which of these theories is most likely to be correct.

A bizarre fossil reptile from the Late Triassic of Russia, called *Sharovipteryx*, is thought by some to be part of pterosaur ancestry. Like pterosaurs, it has thin wing membranes apparently supported internally by flexible rodlike structures. However, in *Sharovipteryx* the forelimbs are tiny, the hindlimbs are tremendously elongated, and there are huge membranes stretching from each hindleg to the tail. This is almost exactly the reverse of what is seen in pterosaurs. Thus if *Sharovipteryx* is anything to do with the origins of pterosaurs, it is well off the main line of their evolution.

Exceptionally well-preserved specimens also show that pterosaurs were furry, with hairlike structures covering most of their bodies and even some parts of their wing membranes. As in furry insects and mammals, fur in pterosaurs may have evolved to retain body heat and might show that pterosaurs could generate their own heat internally.

How do we know what Triassic pterosaurs ate?

Early pterosaurs were probably generalist predators that caught insects and small, backboned animals like lizards and fish. *Eudimorphodon,* first discovered in 1973 in Italy but now known from North America and Greenland as well, has conical pointed teeth at the front of its beak and numerous multi-pointed teeth lining the back of its jaws. These appear well suited for gripping and crushing small fish. Further examination has revealed the scales of a small fish preserved in the stomach region of the original specimen. One specimen of *Preondactylus,* a Triassic pterosaur from Italy, is preserved as a tightly packed but jumbled mass of bones which look similar to a regurgitated pellet, of the kind spat out by certain predatory fish. The implication is that this unfortunate pterosaur was captured and eaten by such a fish, perhaps when hunting over the water. These finds suggest that, early on in their history, pterosaurs became committed to a life tied to aquatic resources. Consequently, many of the later forms were destined to become predators of aquatic animals.

Peteinosaurus in *Walking with Dinosaurs* is also from the Late Triassic of Italy and is a diminutive pterosaur with a wingspan of perhaps two feet. It appears to be related to *Dimorphodon,* a better-known, larger form from the Early Jurassic of England and North America. Presumably, like *Dimorphodon, Peteinosaurus* had a deep skull with a curving upper edge to its snout. At times these skulls have been likened to those of modern puffins, which use their deep bills as colorful display structures and also to catch and hold fish. The numerous small teeth of *Peteinosaurus* suggest, however, that it was an insect eater. It is unlikely that early pterosaurs like *Peteinosaurus* were able to catch very fast-moving prey, like dragonflies. Such insects have changed little from their appearance in the Triassic, and living dragonflies are so nimble in the air that they are immune to most predators, even flying ones. Only certain species of falcon routinely catch dragonflies, and falcons are among the fastest flying of all birds.

Global change at the end of the Triassic

By the very end of the Triassic, the dicynodonts had all but gone. Giant predatory archosaurs like the phytosaurs and *Postosuchus* were also almost extinct. A controversial fossil from 50-million-year-old rocks, called *Chronoperates*, suggests that primitive cynodonts might have survived throughout the whole of the Mesozoic, but otherwise such animals became extinct in the Jurassic. Mammals, which evolved from the early cynodonts in the Late Triassic, did become increasingly successful after the Triassic, but only as small, mostly burrow-dwelling animals. Life on land, at this point in time, was to be dominated by the dinosaurs.

Fossil assemblages from the latest Triassic show that theropods like *Coelophysis* were now the sole large predators. The beaked dinosaurs called ornithischians, which later on would include *Iguanodon*, the armor-plated stegosaurs and other forms, now appeared as low-browsing plant eaters. Giant, high-browsing herbivores also appeared in the form of the prosauropods, long-necked giants related to the gigantic sauropods of the Jurassic and Cretaceous. Prosauropods were built to reach up and eat leaves from trees, and were the first herbivorous animals to exploit treetop foliage.

Plateosaurus – one of the first giant dinosaurs

▷ Among the earliest of giant plant-eating animals, *Plateosaurus* was a long-necked prosauropod dinosaur. It had a distinctive curving snout and is known from hundreds of specimens found at sites in Germany and elsewhere.

This major Late Triassic shift in the dinosaurs' fortunes was depicted in *Walking with Dinosaurs* by the appearance of *Plateosaurus*, one of the largest prosauropods. Known from Europe, *Plateosaurus* was a giant of 26 feet and perhaps three tons, with a distinctive down-turned tip to the end of its snout. Its body and limbs were stout – its robust arms indicate that it could walk on all fours, though it could probably also walk slowly on two legs. *Plateosaurus* was first named in 1837, making it one of the earliest named dinosaurs.

In trying to visualize dinosaurs in living form, paleontologists look for the distinctive marks made by muscles and compare these with the muscles of living animals. One discovery made in this way about the anatomy of *Plateosaurus* was that it had a remarkably long and muscular tail. This would have served well as a counterbalance to the

body and neck and might also have been used as a prop when the animal was standing on its back legs. A particularly controversial idea, first proposed by maverick paleontologist Dr. Robert Bakker in 1986, is that plateosaurs could actually lean back on their tails and kick out quickly with both hindlimbs when fighting. Kangaroos do this today, but they don't weigh three tons!

Bendy backs and bent knees: reconstructing *Plateosaurus* for *Walking with Dinosaurs*

Plateosaurus is shown in *Walking with Dinosaurs* walking at a sedate pace. This seems reasonable: among modern animals even those that can run fast spend the vast majority of their lives moving at normal walking speed, so we can be confident that this was also the case for dinosaurs. Though it was not shown doing so in *Walking with Dinosaurs*, the question arises as to whether *Plateosaurus* was capable of moving at speed. Again, reconstruction of this dinosaur's anatomy from fossils shows that it had a long and mobile back. This appears to have allowed considerable up and down movement of the part of the back in front of the hips, suggesting to some that plateosaurs ran with a sort of bounding gallop.

In *Walking with Dinosaurs, Plateosaurus* is shown with hindlimbs that have strongly bent knees. This looks somewhat odd in an animal of this size, as we are used to seeing living elephants – the largest land animals around – and they have straight legs. The fossils show, however, that unlike elephants, which have columnlike legs clearly designed to support weight, *Plateosaurus* and other prosauropods still had knee and ankle joints that were held in a bent, or flexed, position. We can be confident that this was the case in life, because if the fossils are manipulated so that the joints are made to be straight, as they are in elephants, they become disarticulated. *Plateosaurus* and most other dinosaurs therefore seem to have been like living birds, which have special tendons and joints allowing their knees and ankles to stay in flexed postures.

The teeth of *Plateosaurus* had large crowns with coarsely serrated edges. Coupled with its abundance at some sites and, as the shape of its ribcage indicates, the huge hindgut it possessed, it was clearly a herbivore that processed poor quality vegetation. At least some prosauropods swallowed stones that they used to help break down plant material – a method of digestion seen in the sauropods and also in living herbivorous birds. Like most prosauropods, *Plateosaurus* had a huge recurved claw on its thumb. Maybe this was used in self-defense.

Did *Plateosaurus* live in herds?

The plateosaurs in *Walking with Dinosaurs* move together in a huge herd. The idea that *Plateosaurus* might have done this comes from the association of tens of skeletons at Trossingen, in Germany. This might represent the mass death of a herd in a flash flood, though other suggestions have been made. The animals could have died in drought conditions, or perhaps the fossil "herds" actually represent accumulations of individuals that died over a period of many years.

Walking with Dinosaurs and the paleontologist's view of the Triassic world

The *New Blood* episode symbolizes the changes that occurred in animal communities during the Late Triassic. At the beginning of the episode, the giant archosaur *Postosuchus* is the dominant predator, but is later replaced by carnivorous dinosaurs. Meanwhile, *Placerias*, the last of the dicynodonts, dwindles in numbers until it is replaced ecologically by herbivorous dinosaurs. The first mammallike cynodonts and the first pterosaurs also make their debut in the episode.

Little is known about how many of these animals really lived, and many issues remain controversial. Glimpses are provided by exceptional fossils, such as that of the cynodont *Thrinaxodon* preserved in its sleeping posture and the tooth wear seen in dicynodont skulls. These details allowed *Walking with Dinosaurs* to bring these creatures back to life. This was no easy task as many of the Triassic animals seen in *Walking with Dinosaurs* had never been recreated for television .

The giant dinosaurs that appear at the end of *New Blood* – *Plateosaurus* – truly did herald the shape of things to come. Though *Plateosaurus* was a giant, its relatives the sauropods were to become, amazingly, many times bigger. The rest of the Mesozoic Era was to be dominated by these kinds of dinosaurs and many others.

CHAPTER 2 Time of the Titans

155 MILLION YEARS AGO

The Jurassic was a time of titans. *Time of the Titans*, the second *Walking with Dinosaurs* episode, is set late in the Jurassic period and focuses on the sauropods – the largest animals that ever walked the face of the Earth. During the Late Jurassic, these dinosaurs were abundant and their fossils are well known. One rock unit in particular, the Morrison Formation of western North America, has revealed sauropod bones by the ton. Diverse sauropod types lived here side by side. Giant armour-plated stegosaurs are also known from the Morrison Formation, as are fearsome predators like *Allosaurus*.

Time of the Titans is therefore set at the time when the Morrison Formation rocks were deposited. Details uncovered from there over decades of excavation allowed the researchers behind *Walking with Dinosaurs* to bring a variety of these fantastic beasts to life, and in this chapter we look at the evidence behind these reconstuctions.

Fossils of smaller animals, including pterosaurs and small dinosaurs, are also well-known from these same rocks. Enough information exists to reconstruct the appearances and lifestyles of these animals, so they also star in *Walking with Dinosaurs*. Two, the little pterosaur *Anurognathus* and the theropod dinosaur *Ornitholestes*, were chosen for roles in *Walking with Dinosaurs*.

Introducing the sauropods – the dinosaur titans

Two of the most famous sauropods are *Diplodocus* – long-necked and long-tailed and famous for reaching lengths of 80 feet and more – and *Brachiosaurus*, a giant with elongated front legs and a tall bump on its skull. More work has probably been done on the way these two sauropods moved, fed, and lived than on any other kinds, so it was inevitable that they were chosen to star in *Walking with Dinosaurs*.

◁ **At over 80 feet long and with a tail that could crack like a whip, *Diplodocus* was among the largest of animals to have walked the Earth.**

As is true of all sauropods, the immense size of both *Diplodocus* and *Brachiosaurus* bones is somewhat deceptive: they were actually lightly built, hollowed internally by spaces which, in life, would have housed air-filled sacs connected to the lungs. In *Walking with Dinosaurs*, the sauropods are shown as elephant-like and entirely land-living. This stands in marked contrast to the old idea that sauropods were amphibious – a notion that arose both because it was thought that their legs could not support their immense bulk, and because the Morrison Formation was regarded as a vast swamp environment.

In fact, sauropod legs and bodies are closer in structure to those of land-living large animals like elephants and they lack the features seen in amphibious animals like hippos and crocodiles. The plants and fossil soils of the Morrison are now known to come mostly from dry, semi-arid open woodlands. The land-living, elephant-like sauropods seen in *Walking with Dinosaurs* are therefore backed by firm evidence.

Diplodocus: international star of the sauropods

As one of the world's most famous and most studied sauropods, *Diplodocus* just had to be included in *Walking with Dinosaurs*. How it came to be this famous and well studied is an interesting story. An immensely long but lightly built form, *Diplodocus* was first named in 1878 by the well-known American paleontologist Othniel Charles Marsh. It was first found in Colorado, but it was the discovery of a nearly complete skeleton in Utah that was destined to make a star of *Diplodocus*. Andrew Carnegie, a millionaire who had built his fortune on the steel and railway industries and had established a museum in Pittsburgh, acquired this specimen thanks to its discovery by Earl Douglass, a scientist based at his museum.

Replicas of the skeleton were sent by Carnegie to major museums around the world, among them those in London, Paris, Vienna and Mexico City. Even today they are still the most imposing display pieces in these institutions. These three-dimensional reconstructed skeletons help greatly in visualizing the life appearance of *Diplodocus*, in contrast with those dinosaurs known only from fragments, none of which can be reliably put back together as a skeleton. By examining the shapes and positions of its joints, the attachment areas for its muscles, and then by adding information from fossil skin impressions, scientists have been able to build up a reliable picture of what *Diplodocus* looked like when alive.

△ Parallel lines of sauropod tracks, such as these from Late Jurassic rocks of Colorado, show that sauropods moved in herds. The tracks also show that sauropods did not drag their tails.

▷ Like elephants, *Diplodocus* and other sauropods would have had a tremendous impact on their environment. In eating young trees and pushing over larger ones, they would have created open parkland environments.

Why does *Diplodocus* in *Walking with Dinosaurs* have a spiky back?

In *Walking with Dinosaurs, Diplodocus* is shown with a scaly hide and a row of spikelike scales along the length of its neck, back, and tail. Though artists had long put such spiky decorations onto their reconstructions of dinosaurs with only circumstantial evidence, the presence of such structures in sauropods was confirmed in 1992. Paleontological researcher and artist Stephen Czerkas found a *Diplodocus*-like sauropod in Wyoming that clearly preserves tall, triangular spikes running along the top of its tail. For *Diplodocus*-like sauropods at least, therefore, we do have evidence for a spiny ridge along the backbone, something like that seen in living iguana lizards and tuataras. These living reptiles use the spines when displaying to rivals and when courting – males have larger spines than females and are able to erect them when displaying.

The function of the spines in sauropods is unknown. They may have helped in breaking up the animal's outline, and therefore aided camouflage, or they could have served in courtship and intimidation, as they do in the iguanas and tuataras. From the scanty fossil evidence, the structures in the sauropods do not seem to have been mobile.

Sauropod skin from the limbs, tail and side of the body is known from several well preserved fossils, including the *Diplodocus*-like specimen described by Czerkas. This skin is not wrinkled and thick, as it is in elephants, but covered in rounded or polygonal scales that are mostly about 1/2 - 1 inch across.

Because of their immense size, sauropods frequently left abundant footprints when they walked over soft ground. Such tracks are easily recognised because of the distinctive shapes of the for- and hindprints. Prints of the forefeet are characteristically horseshoe-shaped because of the archlike arrangement of the five fingers and the concave hind margin of the palm. Prints of the hindfeet are large, rounded impressions not unlike those of elephants, but with three large, curving claws.

The shape and function of the sauropod tail

In *Walking with Dinosaurs*, the sauropods have erect tails that project straight out behind the body. Old reconstructions of sauropods usually show them with the tails dragging along the ground. However, sauropod tracks never preserve drag marks made by tails, and articulated sauropod skeletons show that the tail did indeed project horizontally from the hips. In keeping the tail horizontal like this, sauropods opened up a variety of possible uses for it, and one use in particular seems to have been employed by *Diplodocus* and its relatives. The tremendously elongated ends of their tails were used as whips, either to make signals used in communicating with other sauropods, or as weapons.

The bones that make up the end of the tail whip in *Diplodocus* are not particularly strong, however, and some paleontologists think that they would not have lasted long if used to whack predators. In 1989, Professor McNeil Alexander came up with the intriguing idea that whip-tailed sauropods used their tail whips as noise making devices. When swung at low speeds, the *Diplodocus* tail would probably have made a swishing noise, but if swung at high speed, the result might be a loud crack, just like that made by a whip when it is cracked by a human hand. A recent analysis of tail function by Dr. Nathan Myhrvold and Dr. Philip Currie suggests that *Diplodocus* had the muscular strength, freedom of movement and amount of taper in the end of the tail to allow it to swing its tail fast enough to break the sound barrier – that is, the end of the tail would have traveled at more than 700 miles per hour! Maybe, therefore, whip-tailed sauropods used loud cracks from their tail whips to intimidate predators and to communicate with other sauropods. This idea is controversial and highly speculative, so *Walking with Dinosaurs* advisors decided not to incorporate it in the reconstruction.

Could *Diplodocus* have stood on its hindlegs?

As shown by their tracks and the form of their skeletons, sauropods walked on all fours. However, in *Walking with Dinosaurs* the *Diplodocus* is shown rearing up onto its hindlimbs when fighting or pushing over trees. How confident can paleontologists be that *Diplodocus* was really able to do this? Unlike most kinds of sauropod, *Diplodocus* and its relatives have notably short forelimbs, a short thorax (chest region) and peculiarly boat-shaped chevron bones at a point about a third of the way along the length of the tail. These features suggested to the sauropod authority Elmer Riggs in 1904 that *Diplodocus* might have been in the habit of rearing up onto its hindlegs while using its tail as a third 'leg', much like a kangaroo uses its tail as a prop when rearing up. The boat-shaped chevrons could have helped protect the blood vessels running along the underside of the tail, while the short thorax and forelimbs would have required less muscular effort to raise. Studies on the strength of sauropod hindlimb bones by Professor Alexander show that these animals certainly had enough bone strength to bear their weight when standing on their hindlegs. Perhaps, therefore, *Diplodocus* and related sauropods reared up when fighting or feeding. On the basis of this evidence, *Diplodocus* is shown rearing on its hindlegs in *Walking with Dinosaurs.*

A problem regarding bipedal postures in sauropods, however, is that standing on the hindlegs raises the head significantly higher than the heart. In giraffes, where the brain is about 6 feet higher than the heart, the arteries are thickened and have strong elastic walls, and the blood pressure is high. If a sauropod like *Diplodocus* were to stand on its hindlegs, the brain could be as much as 20 or more feet higher than the heart. The blood pressure required to sustain the sauropod brain while standing bipedally would presumably have been exceedingly high. On the basis of these arguments, some physiologists maintain that bipedal postures in sauropods were effectively impossible, or at least sustainable for only very short periods. However, we can be almost certain that male sauropods adopted a bipedal posture while mating!

What do paleontologists know about sauropod eggs and babies?

Walking with Dinosaurs showed a female *Diplodocus* laying a large clutch of eggs in a nest. Later, the eggs hatched and the babies hid in the depths of a Jurassic forest until they became much larger. Perhaps surprisingly, the fossil record reveals many clues as to the nesting strategies of these giant dinosaurs. *Walking with Dinosaurs* pieced together these various lines of evidence to depict one possible interpretation of sauropod life history.

Just as the bones of sauropods are abundant and well represented in the fossil record, so, too, are their eggs. Literally hundreds of examples, including many complete ones, are known from Jurassic and Cretaceous rocks around the world. In many cases, such eggs are

◁ *Diplodocus* has a sliding jaw mechanism and unusual types of wear on its tooth tips. These suggest the "bite and pull" feeding style reenacted for *Walking with Dinosaurs.*

thought to have belonged to sauropods simply because sauropod bones are commonly found in the same deposits. Consequently, there is now some doubt as to whether or not these eggs really were laid by sauropods. Recent finds in North and South America, however, include the association of definite sauropod embryos with eggs and eggshells. Because these eggshells have a distinctive texture when viewed under the microscope, experts have been able to identify with certainty other sauropod eggs found without their embryos preserved.

Eggs thought to have been laid by sauropods are large, about 10 inches across, and spherical. Eggs of this form were therefore reconstructed for the egg-laying scene in *Walking with Dinosaurs.* Sauropod nests appear to have been quite variable in shape, perhaps because different kinds of sauropod made nests of different shapes. Some nests, such as those from Late Cretaceous Spain, are arranged in large arcs, suggesting that the females moved in a semi-circle while laying. Others are arranged in lines. Presumably the female sauropods squatted down when laying these eggs, so that they did not break.

Sauropod nests typically contain about 40 eggs, but exactly how many eggs a female sauropod would have laid in total remains unknown. In view of their huge size, and the fact that large living reptiles like leatherback turtles and certain snakes produce hundreds of eggs (or, in the case of the snakes, hundreds of live babies), it might be that sauropod females also produced hundreds of eggs. This implies hundreds of babies, and fossil track evidence compiled by Dr. Martin Lockley, the world's foremost expert on dinosaur tracks, indicates that baby and juvenile sauropods made up much of the population by number. Presumably mortality among these babies was high as they would have been vulnerable to predators. Babies are known for *Camarasaurus* from North America and for an unidentified titanosaur species from South America. Both were blunt-headed sauropods somewhat similar to *Brachiosaurus*. These specimens and others suggest that hatchling sauropods were slightly over three feet in length.

Did sauropods look after their babies?

This brings us to the subject of parental care. In *Walking with Dinosaurs*, *Diplodocus* does not exhibit any parental care once the eggs are laid. Crocodiles and birds, the living relatives of dinosaurs, do exhibit parental care, and there is ample evidence from the nests of hadrosaurs and others that dinosaurs may also have looked after their babies once they hatched. Some sauropod tracks do show the footprints of juvenile animals near to those of adults, and occasionally the bones of juveniles are also found adjacent to those of adults. However, evidence that hatchlings were cared for by adults is currently lacking. It might be that baby sauropods were like modern-day scrub fowl, birds related to chickens that incubate their eggs in large mounds of warm soil or decomposing vegetation, and whose chicks are independent upon climbing out of their eggs. In the absence of any better evidence, this strategy was adopted for the *Diplodocus* babies in *Walking with Dinosaurs*.

How do we know what sauropods ate?

Owing to their small size, baby sauropods surely fed on low-growing plants, perhaps ferns. This probable behavior was reconstructed for *Walking with Dinosaurs*. A distinctive feeding style was reconstructed

SAUROPOD DIET AND DIGESTION

NO SAUROPOD'S STOMACH contents have yet been found, but there has been plenty of speculation on how they fed and what they ate. There is good evidence that they were herbivores – the sheer size of the animals and the shape of their teeth indicate a herbivorous diet, and biomechanical analyses of their skeletons show how they cropped leaves. Anatomical evidence shows that nearly all sauropods had rather limited jaw movement and little processing of food would have occurred in the mouth – unlike *Iguanodon* and its relatives (see chapter 4) sauropods did not chew their food.

Direct evidence for the mode of digestion used by sauropods comes from specimens whose stomachs are full of smoothly polished stones called gastroliths. Many living birds swallow stones or grit to help grind up their food, and it appears that sauropods did the same. Their huge size indicates that they probably had enormous guts where plant food would have been further broken down by abundant gut bacteria. Living animals that use this method, called hindgut fermentation, produce large amounts of methane.

Material thought to represent sauropod feces contains abundant plant debris. Cretaceous dinosaur coprolites (dung piles) from Montana preserve filled-in burrows made by dung beetles, showing that by this time these insects had evolved their specialized dung-eating lifestyle. Such processing by insects might in part explain the relative rarity of fossilized dinosaur dung.

for *Diplodocus*: when eating ferns, the young animals were shown grabbing the fronds and then pulling sharply upwards, stripping the fern of its leaves. This idea comes from investigation of wear seen on the tooth tips of *Diplodocus* by Dr. Paul Barrett and Dr. Paul Upchurch. Prior to their work, no one had really appreciated how odd the feeding style of *Diplodocus* really was.

Though the juvenile *Diplodocus* in *Walking with Dinosaurs* are shown feeding on ferns, what adults ate might be revealed by investigation of their tooth structure, an area that has been the subject of much recent research. Though the teeth of *Diplodocus* are pencil-shaped and appear rather weak, they were arranged in closely fitting rows and clearly had the capacity to crop tough vegetation.

Microscopic analysis of the surface of the tooth enamel of *Diplodocus* and other sauropods by Dr. Anthony Fiorillo shows that *Diplodocus* teeth did not have much scratching on their surface. This suggests that the plants *Diplodocus* was eating were not contaminated by fine silt, as is often true of low-growing plants like ferns that get covered in dust from splashing raindrops. Perhaps, instead, *Diplodocus* ate from taller tree ferns or conifers.

In contrast to *Diplodocus*, a variety of short-skulled sauropods, including *Brachiosaurus* and *Camarasaurus*, had robust, spoon-shaped teeth that appear well suited for tearing leaves and small branches from trees. Fiorillo found that the tooth enamel in these

△ **Examination of wear marks on their teeth suggests that *Diplodocus* fed at high levels, perhaps from tree ferns or conifers.**

sauropods bears much fine scratching indicative of a diet in which some hard silt grains occur. This might be taken to suggest that they were feeding on vegetation at low levels. However, this seems at odds with our understanding of the habits of *Brachiosaurus* – most paleontologists agree that it was a high-level feeder, taking leaves from the tops of trees.

Brachiosaurus and what it looked like

▷ **One of the biggest Jurassic sauropods, *Brachiosaurus* is a long-necked giant known from both eastern Africa and North America.**

A sauropod that features alongside *Diplodocus* in *Walking with Dinosaurs* is the immense giant *Brachiosaurus*, one of the heaviest sauropods known. The body shape of *Brachiosaurus* was quite a contrast to that of *Diplodocus* but, again, well preserved skeletons provided enough information for experts to reconstruct this dinosaur for *Walking with Dinosaurs*. Rather than being long and horizontal with a whip-like end, its tail was shorter and appears to have sloped downwards. The body also seems to have been tilted upwards diagonally, as the shoulders were somewhat taller than the hips. The limbs of *Brachiosaurus* were also especially slim and elongated, resulting in a

body that was particularly high off the ground, as depicted in *Walking with Dinosaurs*.

Two species of *Brachiosaurus* are known, one from eastern Africa, the other from North America. The *Brachiosaurus* portrayed in *Walking with Dinosaurs* is based on *Brachiosaurus brancai*, the East African species. This is because this species is better known than the North American one, even though the North American *Brachiosaurus* was described first. Complete skulls are known for *Brachiosaurus brancai*, and show that it had a broad, rectangular snout and high arching forehead region, but no skull has definitely been found for the North American species. However, a skull very similar to the African ones is known from North America, but unfortunately was not discovered together with any definite brachiosaur bones. If it does belong to one of the American brachiosaurs, it shows that they also have the arched skull crest and rectangular muzzle of the African species.

SAUROPOD NECKS

SAUROPODS ARE FAMOUS for having long necks. While some of the longest-necked forms, such as *Mamenchisaurus* from the Late Jurassic of China, have as many as nineteen vertebrae in their necks, *Diplodocus* has fifteen. These vertebrae are linked together with a ball and socket joint on the main body of the vertebra, but they are also linked at several other places. Notably, the ribs of these vertebrae are elongate and project backwards down the neck, overlapping the following two or three vertebrae. There are also projections above the main body of the vertebrae, called the zygapophyses, which stabilize the neck. The overlapping ribs apparently reduce movement, although it is probable that the ribs themselves were not completely rigid – bone can be quite flexible.

By constructing computer-generated three-dimensional models of sauropod neck vertebrae, Professor Kent Stevens and Dr. J. Michael Parrish attempted to determine the degree of flexibility in the necks of *Diplodocus* and its more heavily built relative *Apatosaurus*. By analyzing the necks on computer, Stevens and Parrish were freed from the constraints of manipulating real fossils, which are frequently crushed, distorted, and too big and heavy to handle. To determine how soft tissue structures such as muscles and ligaments limited the degree of motion, they also examined neck mobility in birds. They found that for the sauropods, neck mobility was restricted and the necks were essentially horizontal. When standing on all fours, *Diplodocus* had a maximum vertical reach of 13 feet, thus it was barely able to feed at levels higher than its own back. As *Diplodocus* moved its neck sideways, the head was forced slightly upwards. *Apatosaurus* proved more flexible, being able to reach the ground as well as further upwards with ease. Also, when turning its neck to the side, the head of *Apatosaurus* was not forced upwards as it was in *Diplodocus*. Thus *Apatosaurus*, when standing on all fours, had a bigger feeding range than *Diplodocus*.

Stevens and Parrish argued that these limitations to movement in the neck indicated that ground feeding was the primary mode of food gathering for these animals. However, these conclusions are contradicted by studies of tooth wear and are regarded as controversial by other paleontologists, some of whom point out that the degree of motion seen in neck bones does not always reflect the degree of motion available during life.

Armor plates and spiky tails – the stegosaurs

Sauropods were not the only giant herbivores of the Jurassic forests and fern prairies. *Stegosaurus* was a huge, striking dinosaur that reached 23 feet in length and three tons or so in weight. This imposing dinosaur also makes an appearance in *Walking with Dinosaurs.* Though not the largest stegosaur (*Dacentrurus* from Europe reached 3 feet), *Stegosaurus* was among the largest and certainly had the largest plates. It is also the best-known stegosaur, as several complete individuals have been found, all from the Morrison Formation

The stegosaurs that star in *Walking with Dinosaurs* make full use of their plates, using them in display and combat. These amazing structures have long been a source of fascination to paleontologists, and the way *Stegosaurus* uses them in *Walking with Dinosaurs* reflects some of the latest thoughts on their function. While the plates of *Stegosaurus* were made of bone, they would have been covered in life by skin and probably by a keratinous or scaly sheath. Internally, the plates have a rather spongy texture, the spaces of which would have been occupied by blood vessels. There are no sites for muscle anchorage at the bases of the plates, showing that they were immobile.

What were the plates for?

The function of the plates in *Stegosaurus* has been a subject of continuous debate. They might have been used as armor, as display structures for courtship, or to regulate body temperature. Of course, they could have been multifunctional structures used for all or some of these purposes. Wind tunnel studies on the diamond-shaped plates show that they were ideally suited for heat exchange, providing that they were arranged in an upright, alternating pattern. A recent idea proposed by Dr. Kenneth Carpenter is that *Stegosaurus* might have made itself look even more intimidating during combat by "blushing" – that is, by pumping blood into the skin over the plates and thereby making them flush pink or red. This new idea was incorporated into *Walking with Dinosaurs.*

How the plates were arranged in life has also been controversial. The stegosaurs in *Walking with Dinosaurs* have them arranged in two rows that are not symmetrical but have the plates in an alternating sequence. However, early suggestions include that they were arranged like tiles over the animal's sides, or in a single, continuous row along

△ With its short feet and column-shaped legs, *Stegosaurus* was not built for speed. The bony armor on its throat, back, and tail, highly evident in fossil skeletons, probably functioned in self defence.

▽ Complete skeletons prove that *Stegosaurus* did have its diamond-shaped plates arranged in two alternating rows. Perhaps these were used to make the animal look more intimidating, or attractive.

the midline of the back. Paleontologist Frederick Lucas argued in the early 1900s that they were arranged in two alternating rows either side of the midline, as in the most complete specimens there were not enough plates for two symmetrical rows. A well preserved, complete specimen discovered in Colorado in the late 1800s, and another found by researcher Brian Small in the 1990s, provide the definitive answer. They were indeed arranged in two alternating rows, as seen in *Walking with Dinosaurs.*

Ironically, it now appears that *Stegosaurus* was an exception among the stegosaurs – all the other kinds seem to have had plates that were arranged in symmetrical pairs. *Stegosaurus* was also unusual in lacking spikes on its shoulders, which all other stegosaurs had.

THE SIZE OF SAUROPODS

THE IMMENSE SIZE of some dinosaurs has always been a source of fascination, for scientists as well as for the general public. Not only did sauropods attain enormous lengths, some forms were also incredibly heavy. However, the heaviest were not necessarily the longest. The Late Jurassic giant *Brachiosaurus* may have weighed about 77 tons – by comparison, the largest recorded African elephant, shot in Angola in 1955, weighed in at around ten tonnes. What might have been the heaviest sauropod is *Argentinosaurus*, a titanosaur from the Cretaceous of South America, estimated by expert Dr Dale Russell to have weighed more than 110 tons.

Right from the very first sauropod discovery, that of the Jurassic English sauropod *Cetiosaurus*, it was clear that this group of dinosaurs were huge. Sir Richard Owen, the scientist who described *Cetiosaurus* in 1841, estimated that it measured approximately 43 feet in length. Later discoveries of near complete sauropod skeletons in North America showed for the first time, not only their true body shape, but also that they reached lengths of 82 feet in the case of *Diplodocus*.

Much later, discoveries of fragmentary skeletons showed that some sauropods reached even greater sizes than this. *Supersaurus*, discovered in 1972 by James Jensen in the famous Dry Mesa Quarry in Utah, appears to have been similar to *Diplodocus* and is estimated to have reached a maximum length of 150 feet. An apparently even larger relative of *Diplodocus*, dubbed *Seismosaurus* (which means "earthquake lizard"), was discovered in New Mexico and named by Dr David Gillette in 1991. Gillette estimated this creature to be between 128 and 170 feet long! However, more recent estimates have scaled *Seismosaurus* down to about 100 feet, meaning that *Supersaurus* was actually somewhat bigger. It also appears that the one known *Seismosaurus* individual is not, in fact, truly distinct from *Diplodocus* and represents an old, possibly fully grown *Diplodocus*.

Sauropods appear to have grown continually throughout life. Although their growth rate probably slowed during later life, this still meant that the oldest individuals were the longest and the heaviest.

What might be the longest sauropod of them all was actually discovered near the end of the nineteenth century. Unfortunately, this animal, named *Amphicoelias fragillimus* by the famous American paleontologist Edward Drinker Cope, is known only from a single broken vertebra, and this never seems to have been collected from the place in which it was discovered! Based on drawings of this vertebra made in the quarry, dinosaur expert Greg Paul calculated a possible maximum size of 200 feet and 165 tons for *Amphicoelias* – thus it was possibly even larger than the blue whale, the largest living animal.

The stegosaur tail: a formidable weapon

In *Walking with Dinosaurs*, we see *Stegosaurus* wielding its spiky tail tip as a truly formidable weapon. Such a function seems highly likely but, as with the plates, the orientation of the tail spikes has been somewhat problematic. Old reconstructions show these spikes pointing upwards but, positioned in this way, they do not fit onto the tail bones because of their expanded bases. Positioned correctly, they point sideways, backwards, and only slightly upwards. The large size and elongated, pointed shape of these spikes strongly suggests that they were used in defence, perhaps in defending *Stegosaurus* from *Allosaurus* and other large theropods. Verification that *Stegosaurus* used its tail spikes for defence comes from the amazing new specimen discovered by Brian Small. In this animal, one of the formidable tail spikes is deformed, apparently by injury and subsequent infection.

Giant predators from the time of the titans

Sauropods and stegosaurs may have been well protected from predatory dinosaurs, both by their large size and their whiplike or spiky tails. But they were almost certainly not invulnerable, especially young or infirm individuals. During the Late Jurassic, various giant predators inhabited western North America, and appear well equipped for hunting and killing giant dinosaurs. *Ceratosaurus* is peculiar in having horns on the end of its snout and over its eyes, as well as exceptionally elongated teeth in its upper jaw. *Torvosaurus* had no horns and smaller teeth, but its arms were tremendously stout and powerful and its three hand claws were huge, sharp, and strongly curved.

Allosaurus – lion of the Jurassic

Allosaurus, which appears in *Walking with Dinosaurs*, was the most common of these giant carnivores and, from an ecological point of view, perhaps one of the most important. Numerous skeletons of this large theropod are known from Late Jurassic rocks of Wyoming and Utah, and in 1998 *Allosaurus* was also discovered in Late Jurassic rocks of Portugal. This discovery makes *Allosaurus* one of several dinosaurs known to have occurred in both North America and Europe at the end of the Jurassic – evidence for the land connection that existed between these land masses at this time.

In many ways, *Allosaurus* is peculiar for a large theropod. Unlike most others, its teeth are proportionally small and it also has a unique bone in its lower jaw, set just in front of the jaw articulation. It also lacks a remnant of a fourth finger, which all of its close relatives had. The skull of *Allosaurus* is very narrow and there are erect triangular horns in front of the eyes. In *Walking with Dinosaurs*, these horns were depicted as being bright red. Why? It is likely that these horns were used for sexual display – their bright color in *Walking with Dinosaurs* emphasises this possible role. Similar display structures in living animals, such as the head crests of certain birds and the horns and crests of some lizards, are also brightly colored.

The body of *Allosaurus* was rather narrow and its tail was stiffened in its end half. A giant *Allosaurus* specimen from Wyoming is about 40 feet long, indicating that this animal reached almost the same size as *Tyrannosaurus* of the Late Cretaceous. Most other adult individuals of *Allosaurus* are about 33 feet in length.

Was *Allosaurus* a caring parent?

An intriguing site in Wyoming reveals a possible insight into the parental care strategies of *Allosaurus*. Described in 1996 by Dr. Robert Bakker, the site contains sauropod bones, the complete skeleton of an adult *Allosaurus*, and numerous tiny teeth that, Bakker argues, belong to baby *Allosaurus* specimens. The sauropod bones have miniature score marks that match exactly the size and shape of the baby *Allosaurus* teeth from the same site.

Bakker suggests that adult allosaurs brought back pieces of meat for their babies, and that the babies stayed together in a protected den. On this occasion, one of the parents died, explaining the adult skeleton found at the site, but the absence of juvenile skeletons suggests that they survived until the time when they left the den. This discovery suggests that parental care in large predatory dinosaurs was well-developed.

How did *Allosaurus* kill, and what did it eat?

The allosaurs in *Walking with Dinosaurs* are dangerous ambush predators that attack and kill – or try to kill – both juvenile and adult sauropods. Much evidence supports this lifestyle, though many of the

△ *Allosaurus* was an important Late Jurassic predator and had a deep, narrow skull and strong arms. Perhaps the horns over its eyes were for sexual display.

▷ Two of the best known Morrison Formation dinosaurs, *Allosaurus* and *Stegosaurus*, in a confrontation. Direct evidence that *Allosaurus* attacked *Stegosaurus* is lacking, but the two would certainly have been familiar with each other.

details remain open to speculation. The teeth, claws, and body shape of *Allosaurus* show that it was certainly a carnivore, and its large size suggests that it could kill large dinosaurs. How *Allosaurus* might actually have killed, however, remains contentious. It is thought that, like most other predatory archosaurs, these dinosaurs ambushed their prey from cover, since they do not appear to have been built for sustained running. Their recurved, serrated teeth appear best suited for making long slices in the bodies of their prey, and perhaps they used the three sharply curved claws on their hands to make raking wounds as well. Some palaontologists have argued that allosaur skull bones were only weakly connected by ligaments and could bow outwards when the jaws were opened wide, or when the animal swallowed a large chunk of meat.

During *Time of the Titans*, we see two allosaurs moving together as a pair. This reflects another area of speculation: whether *Allosaurus* was a social hunter or not. Some tracks made by large theropods show the animals moving together in groups, so perhaps they hunted in packs. If this were so, they could have cooperated in killing even the largest of sauropods.

We do know that *Allosaurus* fed on sauropods, as some sauropod skeletons bear marks made by *Allosaurus* teeth. Furthermore some sauropod skeletons have been found with *Allosaurus* teeth nearby. Perhaps these teeth broke off while the allosaur was feeding. We cannot be certain that the allosaurs killed these sauropods, but it seems highly likely. Evidence that large theropods related to *Allosaurus* preyed on sauropods comes from a fossil track from the Early Cretaceous of Texas. This trackway shows a large theropod following a *Brachiosaurus*-like sauropod. The theropod appears to change direction shortly after the sauropod trackway also shows a change of direction. Unfortunately, the track is not complete, so we do not know if the two animals really did interact.

The small predator *Ornitholestes*

Small predatory dinosaurs, about the same size as allosaur babies, lurked in the undergrowth of the Morrison world. One of the best-known was *Ornitholestes*, a six-foot-long animal that featured in *Walking with Dinosaurs*. *Ornitholestes* was a member of a group of advanced theropods called coelurosaurs, the group that also includes the tyrannosaurs, dromaeosaurs, and ornithomimids of the Cretaceous. Birds share many unique features with coelurosaurs and appear to have evolved from this group of theropods in the middle of the Jurassic.

Why is the *Ornitholestes* in *Walking with Dinosaurs* shown with quills?

Ornitholestes in *Walking with Dinosaurs* was provided with erectile quill-like structures, something like stiff feathers. The presence of these structures reflects recent discoveries made about the life appearance of such small theropods. The close affinity that coelurosaurs have with birds, and the possibility that they might have been "warm blooded", had previously led to the suggestion that they might have had a covering of featherlike structures, or perhaps even true feathers. Spectacular discoveries made in Early Cretaceous rocks of China appear to confirm these ideas. Here, a variety of small coelurosaurs are preserved with feathers or featherlike structures adorning their bodies. Consequently, the *Ornitholestes* in *Walking with Dinosaurs* was given featherlike quills.

By inference, the other small theropods in *Walking with Dinosaurs* should also have been reconstructed with quills. However, many scientists still say that further work is needed before we can be sure that these structures really were feather-like and not, for example, internal skin fibers of some kind.

How do we know what *Ornitholestes* ate?

In *Walking with Dinosaurs, Ornitholestes* is a nimble, opportunistic predator that preys on lizards, small mammals, and, when they are available, dinosaur eggs and babies. Small coelurosaurs similar to *Ornitholestes* do provide some information on the diet of these animals. *Compsognathus*, from the Solnhofen Limestone of Germany, has a ground-dwelling lizard preserved in its stomach, while a specimen of the Chinese Early Cretaceous coelurosaur *Sinosauropteryx* died shortly after eating a small mammal. These discoveries show that small coelurosaurs like *Ornitholestes* were capable of catching small, sometimes fast-moving, backboned animals.

Ornitholestes is known from a nearly complete skull found associated with a partial skeleton in Morrison Formation rocks in Wyoming.

△ **Perhaps using its tall nose horn as a brightly colored display structure, little *Ornitholestes* was a nimble theropod with three-fingered hands and a deep head.**

An isolated, three-fingered hand with sharp, recurved claws is also thought to have belonged to *Ornitholestes*, though this cannot be confirmed. Some of the lower leg bones in *Ornitholestes* are somewhat shorter than is typical for small coelurosaurs.

Named by Henry Fairfield Osborn, the paleontologist who named *Tyrannosaurus rex*, *Ornitholestes* was originally imagined as a nimble predator that could have preyed on early birds, like *Archaeopteryx*. Reflecting this idea, Osborn's name for this dinosaur means "bird robber." At the time Osborn described *Ornitholestes*, it was thought that *Archaeopteryx*-like birds were present in the Morrison Formation. These specimens have since been shown to be of pterosaurs, though some scrappy Morrison remains might belong to birds after all.

The skull of *Ornitholestes* has a relatively deep lower jaw with a slightly downturned tip. The teeth were short at the front of the snout but further back they were somewhat longer, more bladelike and more recurved. Most of the teeth were restricted to the front half of the skull and were situated well in front of the eye socket. A long overlooked feature of the skull is that it preserves the very base of a horn on the end of the nose, as portrayed in *Walking with Dinosaurs*, but the exact shape and size of this horn is not known.

Small pterosaurs from the time of the titans

Besides small dinosaurs, the Late Jurassic world was also inhabited by many miniature pterosaurs, only a few of which had a wingspan of larger than about three feet – a marked contrast to the evolution of giant pterosaurs in the Cretaceous. One such small pterosaur, *Anurognathus,* featured in the *Time of the Titans* episode. Thanks to several localities that preserve pterosaurs with exceptional fidelity, we know much about the soft tissue anatomy of Jurassic pterosaurs and how they might have looked in life.

One such site, Solnhofen in Bavaria, southern Germany, is famous for its fossils of the first bird, *Archaeopteryx,* as well as for pterosaurs. Most of these are of the short-tailed *Pterodactylus* and the long-tailed *Rhamphorhynchus,* well known for the diamond-shaped vane on the end of its tail and its protruding front teeth. Much rarer forms are also known from Solnhofen, such as the tiny, short-skulled *Anurognathus ammoni,* named in 1923 by German paleontologist L. Döderlein from a single specimen preserved mostly as the impression of its skeleton.

◁ ***Anurognathus,*** **a small long-winged pterosaur known only from one specimen, had a deep skull. Its small pointed teeth show that it was an insectivore.**

The life appearance of *Anurognathus*

As shown in *Walking with Dinosaurs, Anurognathus* was tiny, with a wingspan of only 20 inches and a nose-to-tail length of about four inches. Its wings were particularly long and their proportions suggest that the wing membranes were narrow and elongated, a shape suggestive of fast, maneuvrable flight. However, its hindlimbs were also elongated, being nearly as long as the nose-to-tail length, so perhaps the wing membranes were broad adjacent to the body and only narrowed near to the wing tip. As is preserved in some other pterosaurs, there was probably a membrane connecting the hindlimbs together, called the uropatagium. The tail in *Anurognathus* is very short, which is surprising because features of the skull and wing show that it belonged to the rhamphorhynchoid pterosaurs, a group that otherwise all possess elongated tails.

In *Walking with Dinosaurs, Anurognathus* is shown standing on branches and clinging to the sides of sauropods. Obviously we have no evidence that it ever climbed on sauropods, but the shape of its feet indicates that it could have clung to branches and tree trunks. Like other rhamphorhynchoid pterosaurs, *Anurognathus* had an elongate fifth toe (the little toe in humans) that projected sideways from the side of the foot. This could have been used in grasping if *Anurognathus* were to stand among branches.

What did *Anurognathus* eat?

Anurognathus in *Walking with Dinosaurs* is an insect eater that chases the large flying insects that are stirred up by the movement of the giant sauropods. While there is no evidence for any association between *Anurognathus* and sauropods, the idea that *Anurognathus* ate flying insects seems likely on the basis of its anatomy. Its skull was short but deep with relatively broad jaws and its teeth were simple, widely spaced, elongated, and conical. The lower jaw was very shallow. These features all suggest that it was an insectivore, perhaps hunting insects in a similar way to modern long-winged bats and nightjars.

Similar, apparently closely related forms, one called *Batrachognathus* from Russia and another called *Dendrorhynchoides* from China, are also known. They also seem to have had short, deep skulls but in *Batrachognathus* the teeth were strongly curved and more slender. Only four fragmentary remains of this rare group of pterosaurs are known, but at least these show that it was widely distributed and not restricted to the Solnhofen area. For this reason, *Anurognathus* was selected for the *Time of the Titans* episode.

Missing pieces in the Jurassic jigsaw

Time of the Titans was set in perhaps one of the most impressive times in the whole history of life on land, when giant sauropods grew to sizes completely unparalleled by any modern land animals. In reconstructing these animals, *Walking with Dinosaurs* had a special opportunity to depict the animals as huge in size, physically imposing, and at the same time as magnificent and truly awesome. By building on decades of scientific investigation, *Walking with Dinosaurs* clearly succeeded and produced a dynamic Jurassic world populated by giants. However, many questions remain unanswered about this fascinating time.

One thing paleontologists can often be fairly confident about – when the animals are represented by complete, well preserved skeletons – is the way animals looked (except for their color, as discussed in the Introduction). New interpretations can always change things, however, and at the time of writing much excitement surrounds new interpretations of sauropod skull structure by Dr. Larry Witmer, a noted anatomist. Witmer's new reconstructions may mean that older renditions of sauropods will become obsolete.

As is always the case in paleontology, a vast amount of information on the lifestyle of these animals remains unknown. Some evidence does reveal valuable insight, however. Hundreds of fossil sauropod tracks show that the animals moved in herds, so their social behavior in *Walking with Dinosaurs* does have scientific justification. Some evidence suggests that the large predator *Allosaurus* hunted in groups, and a new find from Wyoming shows that they kept their young in dens.

However, next to nothing is known about the actual life cycles of some of these animals. Did *Anurognathus* build nests and feed its babies? Did *Stegosaurus* move in groups? How did *Stegosaurus* mate? Questions like these continue to enthrall palaeontologists and inspire them to search for more evidence. The proposals made in *Walking with Dinosaurs* will help stimulate this search.

CHAPTER 3

A Cruel Sea

149 MILLION YEARS AGO

ALTHOUGH *Walking with Dinosaurs* focused on dinosaurs, the dominant land animals of the Mesozoic, dinosaurs were not the only important reptiles of this world. In the seas, diverse and abundant marine reptiles took the roles of modern seals, whales, and dolphins and thrived in a multitude of fantastic forms. *A Cruel Sea* focuses on the marine reptile community of Middle Jurassic Europe, a time when warm, shallow seas covered much of the land surface and huge concentrations of fish, squid, and plankton teemed in fertile waters.

The sediments deposited in this Jurassic sea are exposed today as a rock unit called the Oxford Clay in England, famed for its use in the brick-making industry. The industrial excavations made in the Oxford Clay have been beneficial to paleontology, as many finds have been made in the brick pits of Peterborough and elsewhere. Exceptionally well-preserved Oxford Clay fossils, and over a hundred years of scientific investigation, have resulted in the reconstruction of one of the most complex fossil marine communities known.

Walking with Dinosaurs shows the Oxford Clay sea to be populated by swimming shelled molluscs called ammonites. Their fossils are among the most abundant in the Oxford Clay, as they are in most marine rocks of the Mesozoic Era. The ammonites of *Walking with Dinosaurs* are joined by masses of small fish, jellyfish, and squid. Two main types of marine reptile evolved to feed on these abundant sea creatures and *Walking with Dinosaurs* features examples of both.

Ichthyosaurs, the "fish lizards", were fully aquatic reptiles that had evolved a triangular dorsal fin, a forked tail and paddle-shaped limbs. Though they were fishlike in shape, we know from their limb girdles and the bones in their skulls that they were not fish, but air-breathing reptiles. The Oxford Clay sea was home to one ichthyosaur in particular, *Ophthalmosaurus*, a very abundant and well-known form named for its gigantic eyes. In *Walking with Dinosaurs* we see how this creature lived, reproduced and how it sometimes died.

◁ Shown using the "alternating downstroke" method of locomotion, this huge *Liopleurodon* is pulling its hindflippers downwards while the foreflippers passively drift upwards.

The second major group was the plesiosaurs, a diverse group of predatory reptiles, all of which had two pairs of winglike flippers. Some plesiosaurs had long necks and relatively small heads with slim, pointed teeth. *Walking with Dinosaurs* features *Cryptoclidus*, one of the best-known Jurassic members of this long-necked group. Other plesiosaurs, in contrast, were short-necked and had huge, elongated heads. These are called pliosaurs and *Liopleurodon* is one of the biggest and best-known members of this group. *Walking with Dinosaurs* shows how this animal moved, how it interacted with other members of its kind, and how it killed and ate its prey.

Other predators were also present in the Oxford Clay sea. There were large sharks, represented in *Walking with Dinosaurs* by the bizarre horned *Hybodus*, and several kinds of marine crocodile. Large bony fishes also lived alongside these reptiles and one fish in particular may have exceeded even *Liopleurodon* in size. This was *Leedsichthys*, an immense filter-feeding form with a head 10 feet long and a total length perhaps exceeding 65 feet. Unfortunately, *Leedsichthys* didn't make it into *Walking with Dinosaurs*!

Walking with Dinosaurs also showed the animals that flew above the Oxford Clay sea and those that perhaps lived along the seashore.

Liopleurodon, killer whale of the Jurassic

One of the main characters in *A Cruel Sea* is the immense predator *Liopleurodon*, one of the biggest of the pliosaurs and almost certainly the killer whale of its day. Perhaps approaching 65 feet in length and a weight of nearly 22 tons, and with a 10-foot skull armed with immense dagger-like teeth, *Liopleurodon* exceeded even the largest of predatory dinosaurs and was one of the largest carnivores of all time. It is known to have swum the Jurassic seas of Europe, Russia, and perhaps South America.

How, and what, did *Liopleurodon* kill?

Walking with Dinosaurs depicts *Liopleurodon* as an awesome predator that attacks and kills the ichthyosaur *Ophthalmosaurus* and, even more spectacularly, grabs the dinosaur *Eustreptospondylus* from the seashore. As to the evidence for this, there is no doubt that, as shown in *Walking with Dinosaurs*, *Liopleurodon* killed and ate other marine reptiles. Numerous bones are known that display *Liopleurodon* bite marks and there are also partial or dismembered marine reptile skeletons that look as if they have been the object of a *Liopleurodon* feeding bout. Some plesiosaur specimens preserved in the brick pits of Peterborough are peculiar in that they are fully articulated, yet they lack their paddles, neck, and head. The most reasonable explanation for this is that these specimens represent animals that had been caught and killed by a *Liopleurodon*, and that the predator had then pulled off and eaten the missing parts.

Walking with Dinosaurs reconstructs how *Liopleurodon* might have caught such marine reptile prey. It is shown patrolling the sea floor, but on detecting an ichthyosaur near the surface it rushes upwards, grasping the ichthyosaur in its huge jaws. The *Liopleurodon* then breaks the ichthyosaur carcass down into smaller pieces for easier swallowing. The "rushing upwards" hunting style depicted in *Walking with Dinosaurs* is based largely on analogy with modern marine predators. Both great white sharks and killer whales use the same technique when they see prey silhouetted at the surface. It therefore seems reasonable to suppose that pliosaurs also hunted in this way. The idea that *Liopleurodon* may have grabbed dinosaurs from beaches, however, lacks firm evidence and is based on inference. Noting the frequent presence of dinosaur bones in marine deposits that also yield pliosaurs, paleontologists have wondered whether the occurrence of the two together is more than just a coincidence. Killer whales off the coast

▷ Perhaps reaching 65 feet in length, *Liopleurodon* was an awesome predator that would have dwarfed other Jurassic marine reptiles. The color scheme depicted here is inferred from what we see today on large marine animals.

of South America have learned to rush up the shores of shallow, shelving beaches and grab fur seals that are lounging near the water's edge. Once they have attacked, the whales pivot their bodies and thrash their tails to return to deeper water. One population of bottlenose dolphins also beach themselves to catch fish. Such behavior is not unique to aquatic mammals; crocodiles routinely lunge out of shallow water to grab animals, be they frogs, birds, antelopes, or humans.

Possible direct evidence that pliosaurs did eat dinosaurs comes from a skeleton of a *Pliosaurus*, a close relative of *Liopleurodon*, discovered in Wiltshire, England, in 1980. Two bony scutes, apparently belonging to an armored dinosaur, were discovered associated with

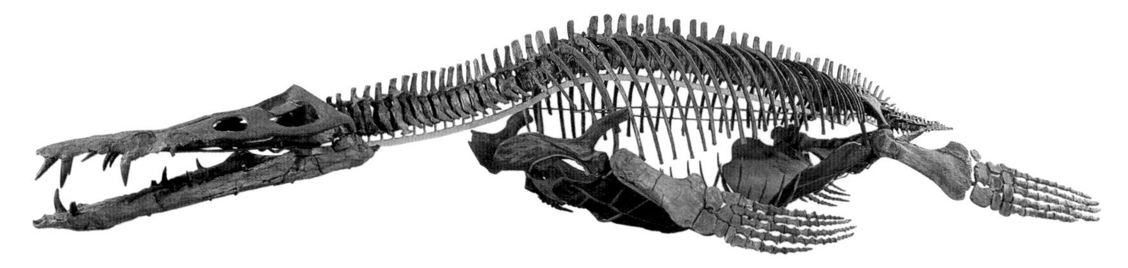

△ The unique four-flippered, streamlined skeletons of *Liopleurodon* and other plesiosaurs appear suited for active underwater flight. *Liopleurodon*'s shortened neck and elongated, toothy skull are clearly evident in this skeleton.

the skeleton of this pliosaur. Though there is no way to be sure, it is likely that these were parts of the pliosaur's stomach contents and therefore show that it had fed on an armored dinosaur. Had the pliosaur killed the dinosaur, or had it merely eaten a floating carcass?

Liopleurodon's underwater sense of smell

Walking with Dinosaurs showed *Liopleurodon* using an underwater sense of smell to detect its prey. Unlike most reptiles, which detect scent particles in air, *Liopleurodon* is shown inhaling water into its nasal cavities and testing this for sensory clues.

A puzzling feature of pliosaur skulls is that their internal nostrils – two holes on the palate – are set further towards the tip of the snout than the external nostrils. Also, the internal nostrils have channel-like furrows that lead into them from the tip of the snout. What was this peculiar configuration for?

Working on a *Rhomaleosaurus* skull, an Early Jurassic English pliosaur, Dr. Arthur Cruickshank, Dr. Philip Small, and Dr. Michael

HOW BIG WAS *LIOPLEURODON*?

LIOPLEURODON IN *Walking with Dinosaurs* is a truly immense creature – the old male that stars in *A Cruel Sea* is a whale-sized giant described as 82 feet long and weighing more than 110 tons. This size created much debate in paleontological circles following the first airing of the program, as no paleontologist thinks *Liopleurodon* really got this big.

Although several complete skeletons have been discovered, these are of individuals of between 16 and 30 feet in length. It is less complete remains discovered in the Oxford Clay that indicate lengths greater than this, though here we move into an area of rough estimates and guesswork. A vertebra at the Peterborough Museum, brought to light in 1996, would seem to indicate a pliosaur of between seventeen and 20 metres in length, and various fragments of snout and lower jaw in other museum collections suggest specimens of similar size. Whether these fragments are actually from *Liopleurodon* is uncertain, and the animal to which they belonged has been nicknamed "*Megapleurodon*." Given that it is unlikely that these bones really do represent the very biggest pliosaur specimens that ever lived, some experts cautiously suggest that *Liopleurodon* and related forms may have achieved total lengths of around 82 feet.

The immense weight ascribed to *Liopleurodon* in *Walking with Dinosaurs* results from comparisons with modern baleen whales. If we accept an imaginary length of 82 feet for *Liopleurodon*, then the only similarly sized living animal is the blue whale.

Because it is not possible to simply put whales onto weighing scales, experts disagree over the weights of these animals. Some say that the largest blue whales may reach an astonishing 220 tones, while others say that they probably don't even reach 110 tons. Regardless, weights within this range were then applied to *Liopleurodon*. However, most of a whale's bulk is carried in the thick blubber layers it carries for use on its long migrations, and to insulate it from the cold of the polar seas it often frequents. *Liopleurodon* was a denizen of warm subtropical seas and would not have had such blubber. We therefore estimate that even the biggest pliosaurs would not have weighed half as much as the biggest whales.

Taylor found the answer. No plesiosaur's mouth provided a watertight seal when the mouth was closed, so when the animal moved forwards, seawater would have entered its mouth. We can infer that, as in crocodiles, a fleshy valve at the back of the mouth stopped water entering the throat. If water was in the mouth, there seemed nothing to stop it from entering the internal nostrils; furthermore, the channel-shaped ducts on the palate would have directed it *into* the internal nostrils.

Cruickshank and his team also found that the internal nasal duct in the pliosaur was an S-shaped tube, able to carry water from the internal nostrils directly to the external nostrils and that the external nostrils had a special backwards-pointing depression. As long as the head was being moved forwards, this would have created an area of negative pressure that sucked water backwards out of the snout chamber. A swimming plesiosaur, therefore, would have had a constant flow-through system of water in its snout! Water would first have been sucked into the internal nostrils, then it would have moved upwards along the internal nasal duct. Here it would have come into contact with the animal's sensitive olfactory organs where information from scent particles would have been passed to the brain. The water would then have been sucked backwards out of the external nostrils.

Though first proposed for *Rhomaleosaurus*, this system now seems to have been true of all plesiosaurs, since all of them have this unusual nostril configuration. Pliosaurs like *Liopleurodon* therefore seem to have combined the hunting strategies of sharks with those of air-breathing predators like crocodiles and whales. Perhaps this system of underwater sniffing was one of the keys to their success, for plesiosaurs similar to *Liopleurodon* were the dominant marine predators for virtually the whole of the Mesozoic Era.

How did *Liopleurodon* swim?

Walking with Dinosaurs shows *Liopleurodon* swimming with an efficient, graceful motion where the two pairs of flippers work in unison. As the front flippers come up, the hind pair comes down, then as the front pair comes down, the hind pair comes up. This system – the alternating downstroke method – ensures perpetual forward motion. As simple as this might appear, it has taken paleontologists decades to work out that this is the most likely model for plesiosaur locomotion.

Early paleontologists thought that plesiosaurs rowed themselves through the water. Rowing, however, is a highly inefficient means of locomotion since no thrust is generated during the recovery stroke. As a result, very few living animals swim by rowing – the commonest examples are insects such as diving beetles and water boatmen.

Most aquatic animals use a swimming mechanism whereby both the downstroke and the upstroke provide forward thrust (or, in the case of the majority of fish, each sideways swipe of the tail fin). This is true for living animals that "fly" underwater, such as penguins and turtles. Plesiosaurs are unlike these animals as they have two pairs of wing-shaped limbs, and they lack the skeletal and muscular features needed to produce a powerful upstroke.

The skeletons of plesiosaurs are remarkable in that all of the main bones for the attachment of the limb muscles are located beneath the body, where they form massive sheetlike structures. These provided very large areas for the attachment of those muscles used in pulling the flippers down, and thus *Liopleurodon* and its relatives would have had an incredibly powerful downstroke. Further proof that these muscles were massive comes from the scars that the muscles left on the limb bones at the points where they were attached.

There is no similarly large area for muscles to perform an equally powerful upstroke. Thus if *Liopleurodon* operated both pairs of limbs at the same time, the effect would have resembled that of rowing – the amount of thrust, though monumental during the downstroke, would have been feeble in upstroke. In resolving this dilemma, paleontologists came up with the ingenious notion that plesiosaurs instead alternated their flipper beat cycle, so while the foreflippers were in the powerful downstroke mode, the hindflippers would be in the weak upstroke mode – so there would always be thrust from at least one pair of limbs. This is the method seen in *Walking with Dinosaurs*, and it works incredibly well.

Smaller marine reptiles, the prey of *Liopleurodon*

Walking with Dinosaurs features two smaller marine reptiles that lived alongside *Liopleurodon* though, presumably, they tried hard to keep

◁ *Cryptoclidus* was a long-necked plesiosaur with numerous small, sharp teeth. It was probably an agile, fast-swimming predator that caught small fish and other sea creatures.

▽ Whether plesiosaurs, like this *Cryptoclidus*, could move on land remains highly controversial.

out of its way! One of these was also a plesiosaur, the long-necked form *Cryptoclidus*. In the series, *Cryptoclidus* is shown twisting and turning through the water, using the same alternating flipper stroke that was reconstructed for *Liopleurodon*.

As shown in *Walking with Dinosaurs*, *Cryptoclidus* was small compared with *Liopleurodon*. Complete skeletons are around 23 feet long. The rather delicate skull of *Cryptoclidus* is shallow-snouted with a deep crest at the back and with enormous eye sockets. The construction of this skull suggests that it was used in catching small, soft-bodied prey – also confirmed by the slim and delicate teeth. One of the most remarkable features of the *Cryptoclidus* group of plesiosaurs is the number of teeth they had: *Cryptoclidus* had nearly a hundred in total and later types, like *Aristonectes* from Chile, had over two hundred. These abundant slim teeth suggest that *Cryptoclidus* fed by using its jaws as a fish trap, engulfing mouthfuls of small prey and straining the water out between its teeth by pushing its tongue upwards.

In the Oxford Clay, *Cryptoclidus* is abundant, being known from hundreds of vertebrae and limb fragments and from individuals ranging from small juveniles to old adults.

Why did plesiosaurs swallow stones?

Walking with Dinosaurs shows *Cryptoclidus* sifting through sand on the seafloor in order to find stones, which it then swallows for use as ballast. We know that plesiosaurs did this because many complete plesiosaur skeletons preserve masses of stones among their stomach contents. In some cases, these stones are not of a local rock type, but come from outcrops many miles distant from where the fossil was found. This suggests that plesiosaurs sometimes traveled great distances to find the right kind of rocks to swallow. Also, the stones found in plesiosaur stomachs are generally of hard rock types, such as quartzite or granite. The animals therefore seem to have preferentially swallowed the hardest stones they could find.

Though it is obvious that plesiosaurs swallowed stones to make themselves heavier, exactly why they would have wanted to do so is a more elusive question. Dr. Michael Taylor, an expert on plesiosaurs, found that living animals that practice "underwater flight", including penguins and sea lions, swallow stones. Taylor noted that swallowed stones were, biologically speaking, less expensive than growing heavier bones. Presumably, underwater fliers find it advantageous to sometimes be negatively buoyant. More recently, Taylor has argued that the more ballast a swimming animal has, the larger its lungs can be without the risk of being positively buoyant. The advantage of larger lungs, of course, is that animals can hold more air and therefore stay underwater longer. So, by swallowing stones, *Cryptoclidus* and other plesiosaurs would have been able to stay near the bottom longer.

Could *Cryptoclidus* move on land?

In *Walking with Dinosaurs*, *Cryptoclidus* is shown resting on land. Whether or not *Cryptoclidus* really did this is the subject of argument among experts. As we have seen, plesiosaurs are peculiar in that their limb girdles are located beneath the animal's body. Paleontologists know this from the many complete, fully articulated plesiosaur skeletons that have been discovered. This configuration would have required tremendous muscular effort if the animal were to try and lift its body when on land.

Plesiosaurs are also unusual in that their limb girdles are not connected to their backbone, as in land animals. This suggests that plesiosaurs would have had great difficulty in transferring force from their flippers to their bodies if they tried to move on land and implies they were built for life in the near weightless conditions of the underwater realm. On the other hand, studies of plesiosaur musculature show that the muscles used to move the flippers downwards were very powerful, so perhaps they were strong enough to lift their bodies up when on land. It is also worth noting that *Cryptoclidus* was lighter than walruses and elephant seals, which *are* capable of moving on land.

Investigations of the sensory abilities of plesiosaurs also suggest that they had evolved for life in the water, and were not built to sense things in the air. We can tell from its skull that, like all plesiosaurs, *Cryptoclidus* had huge eyes. Presumably, it used these to detect its prey and also to keep a lookout for larger predators like *Liopleurodon*. The small bones that would have fitted around the plesiosaur's eyeball (these are called sclerotic ossicles and are found in many reptiles, but not in mammals) show that plesiosaurs had flattened eyeballs well suited for seeing images underwater. These flattened eyes were not designed to cope with the different properties of light in air. Studies of plesiosaur skulls show that their earbones were fused together, meaning that they had also lost the ability to detect airborne sounds properly. As the fossil record shows that plesiosaurs descended from land-living reptiles, they had evidently modified their senses over time to become supremely adapted underwater predators.

Ophthalmosaurus: big-eyed "fish lizard" of the Jurassic seas

The second smaller marine reptile that features in *Walking with Dinosaurs* is the ichthyosaur *Ophthalmosaurus*, a dolphin-shaped squid predator named for its enormous eyes (*Ophthalmosaurus* means "eye lizard"). In terms of relative proportions, these eyes are the largest of any vertebrate animal ever. Some paleontologists think that they are an indication of a deep-diving lifestyle, while others suggest that they imply hunting in murky water. Such large eyes might also have allowed *Ophthalmosaurus* to feed nocturnally, thereby helping to conceal it from large predators that might have hunted using eyesight, like *Liopleurodon*.

Like the other Oxford Clay marine reptiles, *Ophthalmosaurus* is well known and abundant in the rocks of England, Germany, and France but it is now clear that it was also present in the seas that once covered South America, Russia, and elsewhere. A North American ichthyosaur, *Baptanodon*, is regarded by some paleontologists as being the same as *Ophthalmosaurus*, but there is still argument over this. This distribution, and the highly mobile, fast swimming lifestyle we imagine for *Ophthalmosaurus*, suggests it could have lived in Jurassic seas around the globe.

Ophthalmosaurus had a superbly streamlined body. The sharklike dorsal fin and crescent-shaped tail of this ichthyosaur are reconstructed based on fossils that have their soft tissue outlines preserved. The discovery of these fossils makes ichthyosaurs perhaps the best-known group of fossil reptiles, but only in terms of their life appearance. Scientific debate continues over their origin, biology, and their curious extinction – despite being the best-adapted reptiles of all for aquatic life, they did not survive the end of the Cretaceous.

Ophthalmosaurus was between 13 and 16 feet long, although some fragmentary specimens suggest individuals of 20 feet. *Walking with Dinosaurs* shows *Ophthalmosaurus* as having a mouth full of short teeth. Until fairly recently, most experts would have regarded this as inaccurate, as *Ophthalmosaurus* was thought to have been completely toothless when adult. Careful excavations of the adults have since shown that they did have some small teeth set in grooves in their jaws. This near toothless condition might be an adaptation for feeding on squids, like in some living whales. Evidence that *Ophthalmosaurus* was a squid-eater is inferred, and based on stomach contents known for other ichthyosaurs. These show masses of small hooklets that come from the arms of squids.

Reproduction in ichthyosaurs: live birth and bouncing babies

Walking with Dinosaurs reconstructs part of the breeding cycle of *Ophthalmosaurus* as hundreds of pregnant females gather in special shallow areas and give birth to their fully developed, free swimming babies. The babies are shown as being born tail first and fully able to swim from birth. *Walking with Dinosaurs* also shows a mother ophthalmosaur undergoing complications during birth – a misfortune

▽ Baby ichthyosaurs were active and free swimming from birth. Like all baby animals, they had a proportionally larger head and eyes than adults.

◁ Supremely adapted for life in water, *Ophthalmosaurus* was a large eyed ichthyosaur with a shark-shaped body. Its fossils are abundant in the Oxford Clay and many complete skeletons are known.

that spells her doom when a passing *Liopleurodon* takes her as prey.

Our knowledge of Jurassic ichthyosaur biology comes from hundreds of specimens collected at just two or three German localities. Many of these are preserved with their embryos while others are preserved as if in the act of giving birth, with their babies protruding from their cloacas (the external opening for the gut and birth canal in reptiles). That so many pregnant ichthyosaurs are preserved together at select localities suggests that, as *Walking with Dinosaurs* shows, the females moved *en masse* to special protected birthing sites, maybe in shallow lagoons. Specimens preserved with their embryos show that ichthyosaur mothers produced up to eleven babies per litter, though most mothers contain only one or two babies.

Not only do these specimens prove that ichthyosaurs gave birth to live young, they also confirm that, as in whales and dolphins, baby ichthyosaurs were born tail first. However, it is important to note that the individuals "preserved giving birth" were probably *not* giving birth when they died. Instead, they most probably died when pregnant and only later, during decomposition, was the baby pushed out of the birth

THE SOFT TISSUE OF ICHTHYOSAURS

WE OWE THE reconstructed appearance of *Ophthalmosaurus* shown in *Walking with Dinosaurs* to numerous discoveries made in the rocks of England, Germany, and elsewhere. The ichthyosaurs of *Walking with Dinosaurs* are shark-shaped with dorsal fins, forked tails, and two pairs of paddle-shaped limbs. However, prior to the discovery of remarkable ichthyosaur specimens that have their body outlines preserved, nobody knew that ichthyosaurs had such structures. Old pictures show them as being much like crocodiles in their body shape, with straightened tails and no dorsal fins.

German fossils from the Early Jurassic locality of Holzmaden were the first to show that ichthyosaurs had sharklike dorsal fins. Some of these specimens later turned out to be of dubious authenticity (their soft tissue outlines had been augmented by preparators), and some paleontologists consequently suggested that ichthyosaurs might not have had dorsal fins after all. Unlike dolphins, ichthyosaurs retained two rear limbs and could have used these, rather than a dorsal fin, for stabilization. Newer discoveries, however, including some Triassic ichthyosaurs from China, show that some ichthyosaurs at least certainly did possess a dorsal fin. Like in dolphins this was a fleshy structure that had no internal bony skeleton – it is for this reason that it is rarely preserved.

The Holzmaden specimens also show why ichthyosaur vertebral columns are consistently bent downwards at their tail tips, a feature that early paleontologists had regarded as an unnatural break. The tail was actually crescent-shaped, and the tail bones supported the bottom lobe of the crescent. Like the dorsal fin, the upper lobe lacked any bony support, but a tail discovered in Dorset shows that a band of muscle was present here.

Examination of the fossil skin of ichthyosaurs shows that it was smooth, lacked the typical reptilian scales and in some places had a ripplelike pattern. A recent discovery is the presence of stiffening skin fibers arranged in coils around the base of the tail.

canal by gases that built up inside the mother's corpse. This "post mortem birth" is common in beached whales today.

That birth in ichthyosaurs sometimes went wrong is revealed by specimens in which the babies are preserved pointing head first, rather than tail first. Presumably, as is true of whales and dolphins, baby ichthyosaurs were not "triggered" to take their first breath until their head was free of the mother's cloaca. If the baby therefore emerged head first, the possibility of it drowning would be quite high. In a 1995 study on ichthyosaur embryos, Dr. D. Charles Deeming and colleagues suggested that, if this happened, the mother would have been doomed as well, for she then would have been stuck with a dead baby lodged in her cloaca. The decay of the baby would have slowly poisoned her blood stream. Perhaps head first babies did survive sometimes, because successful head first births have been recorded in dolphins.

Another complication that may also have occurred was the abnormal production of too many babies. Live birth is a highly stressful and dangerous process, especially for an aquatic air-breathing animal, and Deeming and colleagues also suggested that multiple births might have killed ichthyosaur mothers.

Horned sharks of the Jurassic seas

Walking with Dinosaurs shows another marine predator of the European Jurassic seas, the large shark *Hybodus*. *Hybodus* is shown as an opportunistic predator and scavenger, and it looks peculiar in having horns on its head and large spines located in front of its dorsal fins. Again, well-preserved specimens of *Hybodus* are known from Jurassic rocks, so we can be confident about the way it looked. Complete specimens show that it had an elongated, torpedo-shaped body, a short, blunt-snouted head, and an elongated tail where the upper lobe was much longer than the lower. *Hybodus* reached ten feet in length.

Hybodus had distinctive multi-cusped teeth with enlarged central cusps. These teeth, together with the small skin teeth or dermal denticles that these sharks had covering their bodies, are among the most common Oxford Clay fossils. Like all sharks, *Hybodus* would have continually shed teeth as new ones grew. *Hybodus* was a particularly successful shark that first appeared in the Triassic and survived until the Cretaceous, living all over the world during this time.

▷ **Probably an opportunistic hunter and scavenger, *Hybodus*, with its two dorsal fin spines, may have followed *Liopleurodon* to pick up scraps after a kill.**

Walking with Dinosaurs shows *Hybodus* following an injured *Liopleurodon*, waiting for it to become moribund. It is also shown as being attracted to an ichthyosaur mother who is experiencing difficulties during birth. As the teeth of *Hybodus* show that it was a carnivore, and as modern sharks are also attracted to wounded or distressed animals, we can be confident that *Hybodus* behaved in a similar way. Furthermore, teeth from *Hybodus* are frequently preserved with marine reptile skeletons, suggesting that, like many modern sharks, it scavenged on carcasses when it found them. All living sharks have an exceptional sense of smell and *Hybodus* almost certainly did too. It would have used this to detect prey. All living members of the shark group also have special sensory pits on their heads that allow them to detect the electromagnetic pulses created by the movements of animals' muscles. *Hybodus* would certainly have had this sense and would have used it to detect prey or danger.

Why all the spikes and horns?

As clearly shown in *Walking with Dinosaurs, Hybodus* had large spines in front of each dorsal fin. In living sharks that have these, like the Port Jackson shark, they serve at self defence, preventing predators (like larger sharks!) from swallowing the animal. The spines may also serve as cutwaters, thereby improving swimming performance, and the fact that their exact shape differs from species to species also suggests that they could have been used in species recognition. With predators as large as *Liopleurodon* around, maybe the fin spines of *Hybodus* proved particularly useful.

Walking with Dinosaurs also shows the large horns *Hybodus* had on the back of its head. It might seem peculiar for a shark to have horns, but some living ratfishes, distant relatives of sharks, have a denticle-covered structure called a head clasper which probably serves the same function. As the name suggests, claspers are used by the males to grip females during mating. The male wraps his body around the female's and uses the head clasper to hold onto the skin on the female's head, thereby helping to keep her still. Modern sharks behave in the same way, but the males lack head claspers. Female sharks have specially thickened skin to help them cope with this kind of treatment but, even so, mating in these animals is not a gentle affair!

Ammonites and kingcrabs

Walking with Dinosaurs features a variety of other Jurassic sea creatures besides the marine reptiles. Among the most abundant marine animals of the Mesozoic were ammonites, a group of swimming molluscs related to squid and octopuses, and in *Walking with Dinosaurs* one kind is harassed by a baby *Ophthalmosaurus.* Ammonites are almost entirely known from their coiled, chambered shells which often had ribbed or spined surfaces. Along with the dinosaurs and most of the marine

◁ **Ammonites were one of the most common animals in the Jurassic seas. Hundreds of different species evolved and many had highly ornamented shells.**

reptiles, ammonites became extinct at the very end of the Cretaceous, but a related group survives today in the form of *Nautilus*, a deepwater scavenger with many tentacles and complex eyes. Fluid- and gas-filled chambers in the *Nautilus* shell allow it to control its buoyancy to a degree, and it swims by jet propulsion by squirting water out of its siphon, a muscular tube through which it breathes.

Though their soft bodies are all but unknown as fossils, it seems reasonable to think that ammonites were similar to this. However, their shell shapes and shell thicknesses show that they were far more diverse than *Nautilus* and included floating, surface dwelling forms, fast moving mid-water predatory forms and sluggish bottom dwellers. Some experts think that ammonites were more closely related to octopuses and squids than to *Nautilus*. Bitten and broken shells show that ammonites fell prey to ichthyosaurs and plesiosaurs, as well as to other kinds of marine reptile and fish.

Invertebrates of a different kind, those with armored skeletons and jointed legs (a group called arthropods), were also abundant in the Jurassic seas. Among them were the ancestors of living crabs and lobsters. *Walking with Dinosaurs* featured another kind of arthropod, a kingcrab, a bizarre bottom-living scavenger with a crescent-shaped armor headshield and a long, spike-shaped tail. Despite their name, kingcrabs are not crabs, but more closely related to spiders. The first kingcrabs appear in the fossil record more than 300 million years ago in rocks of Carboniferous age and they have persisted largely unchanged to the present day, being abundant on the eastern seaboard of the USA and in parts of the Pacific Ocean. In the Jurassic their remains, represented by exquisite complete specimens, have been discovered in the Solnhofen Limestone of Germany.

Long-tailed, fish-eating pterosaurs

Today, seabirds fish from the sea and forage on beaches. During the Jurassic this role was filled by abundant pterosaurs, and among the several forms we meet in *Walking with Dinosaurs* is the toothed, fish-eating *Rhamphorhynchus*, an abundant form that must have been the seagull of its day. *Walking with Dinosaurs* shows *Rhamphorhynchus* as a long-tailed marine predator with a diamond-shaped structure at the end of its tail and wing membranes that are attached to its hindlimbs. Its teeth project forward from its beak, allowing it to grab fish from the surface of the sea, and it is shown skimming along the water surface when hunting, as well as foraging on the shore.

As is true for a number of other pterosaur types, a series of exceptionally well-preserved *Rhamphorhynchus* specimens allowed paleontologists to work out how *Rhamphorhynchus* looked when alive. We know it had a small throat pouch and its wings had narrow, pointed tips but were broader nearer the body where they probably attached to the ankle. The diamond-shaped structure on the end of its tail is preserved in many specimens, and differs in size between individuals. Because it is often asymmetrical it was probably orientated vertically, and because it is proportionally biggest and most strongly asymmetrical in the largest *Rhamphorhynchus* specimens, it was probably a display feature used as a signal to other *Rhamphorhynchus.* We also know (from specimens described as long ago as the 1920s) that, like perhaps all pterosaurs, *Rhamphorhynchus* had body hair.

Like nearly all Jurassic pterosaurs, *Rhamphorhynchus* was small compared with Cretaceous pterosaurs, its wingspan being only three feet.

How do we know what *Rhamphorhynchus* ate?

Walking with Dinosaurs shows *Rhamphorhynchus* as comparable with modern seabirds: a predator that picks fish and other animals up off the water surface but also scavenges along the shoreline and takes other animal prey when it is available. *Rhamphorhynchus* is also shown catching fish and swallowing them head first. Evidence that *Rhamphorhynchus* ate fish comes from the structure of its skull, and from the stomach contents seen in some exceptional specimens from Germany. Several *Rhamphorhynchus* fossils have their stomach contents preserved, and among the food they ate are the remains of small fish, apparently swallowed whole and head first.

Rhamphorhynchus was part of a group of specialized pterosaurs that had their front teeth arranged in a special forward pointing array that called a "fish grab." These teeth were elongated and sharply pointed. Well-preserved *Rhamphorhynchus* specimens show there were sharp beaks on the tips of the upper and lower jaws, and these would also have contributed to the "fish grab." Based on analogy with many modern seabirds, it is thought that *Rhamphorhynchus* would have plucked prey from the water's surface.

The skimming behavior also depicted in *Walking with Dinosaurs* comes from comparisons of the skulls of *Rhamphorhynchus* and of skimmers, a group of gull-like seabirds, by the pterosaur expert Dr. Peter Wellnhofer. Skimmers hunt by trawling their blade-like lower jaws through the water surface and, like them, *Rhamphorhynchus* has a lower jaw that is compressed from side to side.

However, the lower jaw in *Rhamphorhynchus* is broad compared with that of skimmers and is not markedly longer than the upper jaw, as it is in skimmers. Furthermore, *Rhamphorhynchus* lacks attachment sites for the massive skull and neck muscles used by skimmers in absorbing the impact of colliding with objects in the water. These facts cast doubt on the idea that *Rhamphorhynchus* was a skim-feeder, though of course it was not impossible.

▽ The "fish grab" beak of *Rhamphorhynchus* appears supremely adapted for the grabbing and handling of small fish.

▷ With its long tail and toothy, upturned beak, *Rhamphorhynchus* was a common Jurassic pterosaur. Exactly how *Rhamphorhynchus* moved when on land is unknown.

Eustreptospondylus – a beach-combing dinosaur?

As proven by their fossils, many recovered from the marine rocks of the Oxford Clay, dinosaurs lived along the shores of this Jurassic sea. *Walking with Dinosaurs* shows one in particular, the carnivorous theropod *Eustreptospondylus*. This was a bipedal predator with strong hindlimbs, powerful arms with three-fingered hands, and a large, deep head sporting serrated, recurved teeth. These features are evident in a well-preserved, nearly complete skeleton of this dinosaur displayed at the University of Oxford.

Walking with Dinosaurs reconstructs *Eustreptospondylus* as a beach-

▽ Only known with certainty from one Oxford Clay specimen, *Eustreptospondylus* was a large, deep-skulled predator. Its affinities with other large theropods remain mysterious.

combing scavenger that eats carrion from the shoreline and catches pterosaurs. It is also shown swimming from island to island and falling prey to *Liopleurodon*. These speculations are entirely possible, but they cannot be confirmed because *Eustreptospondylus* is only definitely known from the one specimen and we know nothing of its diet, lifestyle, or habits. We do know that the known specimen was preserved at sea, not only because it was discovered in the marine rocks of the Oxford Clay, but also because it has fossil oysters stuck to its thigh bone. Presumably the carcass was washed out to sea after floating in a river, or after it had died near a beach. It is possible, however, that the occurrence of the skeleton here may reflect a coastal lifestyle, or it might have been grabbed and killed by a *Liopleurodon*.

In *Walking with Dinosaurs*, the *Eustreptospondylus* are shown as being about 16 feet long. The Oxford skeleton is about this size. However, as shown by the fact that the vertebrae have not properly fused together, it was still a juvenile when it died. Adults would have been bigger, perhaps growing to more than 33 feet.

Could *Eustreptospondylus* swim?

Perhaps one of the most curious scenes in *A Cruel Sea* is the "fish eye view" of a swimming *Eustreptospondylus* crossing a short expanse of sea. The dinosaur kicks strongly with its hindlimbs, at the same time paddling with its arms. All modern animals (with the exception of a few humans) can swim, and even those that appear poorly suited for it, such as horses and elephants, are actually very good at it. Also, the animals whose bodies are most like those of theropod dinosaurs, large groundbirds like ostriches and emus, are very powerful swimmers that can cross large bodies of water with ease. We can be confident that *Eustreptospondylus* and other kinds of dinosaur were good swimmers.

There is even possible direct evidence for swimming behavior in theropod dinosaurs from fossil tracks. In 1980, Dr. Walter Coombs described a series of Jurassic tracks that are preserved as scrape marks, apparently made only by the tips of a theropod's claws. Coombs' interpretation was that these marks were made by a swimming theropod in which only the tips of its toes were touching the bottom.

Bringing the Oxford Clay sea back to life

In contrast to many other Mesozoic animal communities, a considerable amount is known about the diets and lifestyles of the animals that lived in the Oxford Clay sea. This is because many of the species are known from hundreds of individuals, many of which reveal stomach contents and other evidence of interaction with other animals. Also, several of the forms, such as the sharks and ammonites, still have close living relatives that surely behave in similar ways. Other Oxford Clay animals, like ichthyosaurs, are superficially much like non-related living animals. This makes it easier to make inferences about the biology of the fossil forms. Contrast this with other fossil animals, like sauropods, which simply have no modern-day counterparts. Finally, marine animals in general have a far better fossil record than land animals because they live, and die, in places where sedimentation is more frequent, and fossilisation more frequent, than it is on land.

Compared with most fossil animals, much is known about reproduction in ichthyosaurs, and studies on the skulls of pliosaurs have led to fascinating speculations about their hunting methods. Exceptional preservation and an abundance of specimens (admittedly not in the Oxford Clay, but in the slightly younger Solnhofen limestones) have also been key in reconstructing the feeding methods and diet of the pterosaur *Rhamphorhynchus*. In contrast, the lifestyle depicted in *Walking with Dinosaurs* for the theropod *Eustreptospondylus* was highly speculative as this animal is only definitely known from one specmen.

Recent and current investigations of Oxford Clay fossils continue to increase the complexity of the community known from this deposit. One recent discovery is of a small, bottom-feeding plesiosaur with heavy, thickened ribs. This would have preyed on various kinds of invertebrate, competed with other plesiosaurs like *Cryptoclidus*, and been preyed upon by predators like *Hybodus* and *Liopleurodon*. Animal communities today are incredibly complicated, with hundreds of different interactions occurring between hundreds of different species. It seems likely that fossil communities had similar levels of complexity, but because of elements that are missing from the fossil record, we rarely get glimpses of what these might have been like. The Oxford Clay is one of a handful of exceptions, and what we know of its complexity resulted in its realistic portrayal for *Walking with Dinosaurs*.

CHAPTER 4

Giant of the Skies

127 MILLION YEARS AGO

No dinosaur history would be complete without considering the kings of the Mesozoic skies, the pterosaurs, and in *Walking with Dinosaurs* we meet six of them. Although pterosaurs are *not* dinosaurs in the strict paleontological sense, many paleontologists now believe that the two groups are closely related, and that pterosaurs certainly seem to belong in the group of reptiles – the archosaurs – of which dinosaurs, birds, and crocodiles are also members. Pterosaurs share with dinosaurs distinctive ankle-bone features, and thus appear to be more closely related to dinosaurs than to crocodile-group archosaurs.

In *Giant of the Skies* the camera follows the immense Early Cretaceous pterosaur *Ornithocheirus* on a 8,400 mile migration from South America to Europe via North America. During this journey, we meet some of the other creatures of the Early Cretaceous world including the herbivorous dinosaur *Iguanodon*, the armor-plated *Polacanthus*, the vicious theropod *Utahraptor*, and the early bird *Iberomesornis*.

◁ Giant wingspans were the norm for Cretaceous pterosaurs, as were bizarre head crests. *Tapejara*, with its sail-shaped crest, reached 13 feet while ornithocheirids may have exceeded 33 feet.

Though the reconstructed migration forms the crux of the story, we are not sure that *Ornithocheirus* did this. Needless to say, migration is difficult to demonstrate from fossils as an animal can only die in one place, and evidence of its previous travels is almost impossible to derive from remains of its bones. However, geochemists are optimistic that chemical signatures may be detected in pterosaur teeth that will

BRAZILIAN BONANZA

In *Walking with Dinosaurs*, *Ornithocheirus* begins its journey in what is now Brazil, and the smaller pterosaur *Tapejara* is also shown living here. This reflects the astounding richness of pterosaur fossils in Brazil, particularly on the flanks of the Araripe Plateau in the arid interior of northeast Brazil. Here there are not one, but two, of the world's most important sources of pterosaurs.

These are the Santana and Crato formations, first discovered by the Bavarian naturalist explorers J B von Spix and C F P von Martius in the 1820s. Spix and Martius found that the slopes of the plateau contained hundreds of fossil fish.

More than 100 years later, Ivor Lewellyn Price, one of the most industrious of Brazil's paleontologists, announced the first discovery of pterosaurs from these deposits. What Price had stumbled on was quite remarkable, for he had located the first of what would later prove to be hundreds of the most perfectly preserved pterosaur skeletons known.

The sedimentary layers of the Chapada do Araripe formed at the bottom of saline lagoons during the Early Cretaceous. These lagoons teemed with fish, and pterosaurs must have gathered in huge numbers to feast on them. Now more than 20 different species of pterosaur have been found at these fossil sites. Some of the examples are so well preserved that details of their wing membranes, claws, beaks, soft tissue head crests, and muscles can be seen. Many different growth stages are represented and the skeletons are often found perfectly articulated and in an uncrushed condition.

reveal their previous distribution, much in the same way that elephant ivory can be identified to a location when captured from poachers.

Ornithocheirus – pterosaur giant of the Early Cretaceous

Ornithocheirus was selected from the many different types of pterosaur partly as it has been known to paleontologists for a long time. However, it is not common and no complete skeletons are known, so it did pose a problem for reconstruction. Nearly complete skeletons of closely related forms are known, however, and these provide the basis for the reconstruction in *Walking with Dinosaurs.* The *Ornithocheirus* family of pterosaurs, the Ornithocheiridae, has been recorded in Europe, South and North America, Asia, Australasia, and, most recently, in Africa. So these large pterosaurs were widely distributed.

The history of the name *Ornithocheirus* is complex. It was given by Harry Govier Seeley, the great nineteenth century authority on pterosaurs, in 1869. However, many of the 36 species described as belonging to this genus were discovered before then. In fact, the earliest was described in 1845 by James Scott Bowerbank, who had discovered, with a wingspan estimated at nearly ten feet, the largest flying prehistoric animal known at that time.

As world pterosaur expert Dr. Peter Wellnhofer states, "the name *Ornithocheirus* is something of a 'waste bin' generic name." Because the original material named *Ornithocheirus* is fragmentary, the genus is difficult to diagnose. There are several pterosaurs that appear closely related to *Ornithocheirus* including *Coloborhynchus* from England, Asia, and Texas, *Criorhynchus* from England and *Tropeognathus* from South America. All these forms are clearly closely related, but do crested forms (like *Coloborhynchus* and *Tropeognathus*) represent distinct types, or males of types originally thought to be crestless (like *Ornithocheirus*)?

This problem of sex recognition in pterosaurs might be resolved soon. Dr. Dino Frey of Karlsruhe Museum, Germany, has collected perfectly preserved pterosaur pelvic girdles. He can recognize two distinct types, one with a closed structure and the other with a very open structure. These differences Frey attributes to gender. The open-structured pelvis is apparently from a female and would have allowed the passage of eggs. Of course, Dr. Frey now needs to find a complete skeleton of a crested pterosaur with a closed pelvis and a crestless pterosaur with an open pelvis to confirm his theory.

Why did *Ornithocheirus* have crests on its beak?

Like the *Ornithocheirus* seen in *Walking with Dinosaurs*, all ornithocheirids had long, pointed beaks sporting pointed teeth that interlaced when the jaws were closed. While the original fossil finds of *Ornithocheirus* were of animals with uncrested beaks, the *Ornithocheirus* in *Walking with Dinosaurs* is depicted as being sexually dimorphic – the males have crested beaks but the females lack crests. This reflects a controversy over whether the crested forms are distinct crested types of ornithocheirid or if, as shown in *Walking with Dinosaurs*, they are simply the males of species whose females were without crests.

Walking with Dinosaurs adopted the view of Dr. David Unwin, a pterosaur expert who has studied this ornithocheirid problem in detail. Unwin argues that crested ornithocheirids are the males, and the crestless ones are females. If this is true, it means that male and female ornithocheirids might have fed in different ways. While it is

PTEROSAUR TEETH

ORNITHOCHEIRUS IN *WALKING with Dinosaurs* has elongated, pointed teeth. In contrast, *Tapejara* is completely toothless. These differences reflect the great diversity in the skulls of pterosaurs. Early pterosaurs from the Late Triassic and Early Jurassic had small, pointed teeth and some even had teeth with multiple points, or had different types in different parts of the mouth, as mammals do. Most Jurassic pterosaurs had teeth that were long, slender, and sharply pointed. A few had needlelike teeth that would have been used as filters for straining small animals out of water. Some Early Cretaceous pterosaurs had short, wide, almost petal-shaped teeth, and in one form they were arranged in a tight arc and probably worked like a "pastry cutter," biting off large chunks of meat.

Ornithocheirids usually had long, slender teeth towards the front of the jaws, but shorter teeth (or no teeth) towards the back of the jaw. Pterosaurs such as *Tapejara, Tupuxuara, Pteranodon,* and *Quetzalcoatlus* had no teeth at all. Teeth, owing to their solid enamel parts, would have added considerable weight to a pterosaur skull. In a large toothed form such as *Ornithocheirus*, they may have constituted most of the skull's weight. Being toothless would therefore have been a great advantage for a flying animal as it would reduce weight significantly.

▷ The elongated beak and sharp teeth of ornithocheirid pterosaurs indicate a fish-eating habit.

△ With its enormous wingspan, *Ornithocheirus* would have been an expert glider, able to cover huge distances without difficulty.

◁ Different *Tapejara* species had variously shaped crests. The species shown (it is as yet unnamed) has a tall, rounded crest. Presumably this was used in display, and was therefore vivid in color.

flying over the ocean, we see the *Ornithocheirus* in *Walking with Dinosaurs* using its crests as special keels that allow it to cut its way through the water to grab fishes and squid. Crestless ornithocheirids could not have done this and must instead have plucked prey from nearer the surface. Not all pterosaur experts agree with Unwin. They argue that the crested ornithocheirids – including the male *Ornithocheirus* that featured in *Walking with Dinosaurs* – really are different from *Ornithocheirus* and should be given different names.

Pterosaur head crests

Several types of pterosaur sported head crests and two forms that did, the male *Ornithocheirus* and the smaller, vaguely puffinlike *Tapejara*, were featured in *Giant of the Skies*. Of all the crested pterosaurs, the most famous is *Pteranodon* with its elongated, backward-projecting bony crest. The Brazilian *Tupuxuara* had a large crest that ran in an arc from the back of its head to the tip of its snout. Some had crests on the front of their beak, as in the Brazilian *Tropeognathus* and the English *Criorhynchus*. As shown in *Walking with Dinosaurs*, *Tapejara* had a crest on top of its head as well as on the front of its beak.

It was recently discovered that in some pterosaurs the bony crest supported an extensive area of soft tissue crest. This is most pronounced in *Tapejara* from the Cretaceous Santana and Crato formations of Brazil. Two species of *Tapejara* are known to have had large soft tissue head crests. *Tapejara imperator* has a crest that sits like a sail on top of its head and projects backwards, while the *Tapejara* that starred in *Walking with Dinosaurs* is in fact a new species that has yet to be given a name. This species of *Tapejara* has a small crest at the front of its lower jaw which may have acted as a stabilizer while it was fishing from the surface of the water, but its large head-crest is most likely to be for species recognition and sexual display. It is not yet known if it was males, females or both sexes that sported head crests.

How big was *Ornithocheirus*?

The *Ornithocheirus* in *Walking with Dinosaurs* has a wingspan of 40 feet and was therefore truly a giant among his kind. Ornithocheirids were, until recently, thought to have only reached wingspans of around 20 feet, but new, fragmentary specimens from Brazil suggest that they may have achieved wingspans of more than 30 feet. As with sauropod dinosaurs and some other reptiles, pterosaurs appear to have continued to grow throughout life, so the oldest individuals would have had the greatest wingspans. Contrast this with modern birds, which reach their maximum adult size in one or perhaps two years, and then remain that size for the rest of their lives.

Since the making of *Walking with Dinosaurs* Dr. David Unwin, one of the scientific advisors for the series, has announced the discovery of a new giant of the skies. He has described bones of a pterosaur even bigger than *Ornithocheirus* and *Quetzalcoatlus* (previously the biggest known pterosaur) from latest Cretaceous rocks of Spain. This new animal is thought to have had a wingspan of over 40 feet.

Estimating pterosaur wingspans

Walking with Dinosaurs gave the large male *Ornithocheirus* the maximum possible wingspan of 40 feet. Seeing as such giants are hardly ever discovered as complete skeletons, this raises the question "how do scientists estimate the wingspans of pterosaurs?" While ratios known from complete specimens are combined with attempts at reconstructing the actual wing skeletons, much guesswork is also used.

Paleontologist Dr. Dino Frey described a Brazilian Cretaceous pterosaur that had both wings complete. When he added up the length of all the bones of the wing he found a total wingspan of nearly 16 feet. Another Brazilian specimen in which only one wing was complete (except for a small part of the bone at the wingtip) had a length of nearly 10 feet, thus indicating a wingspan of around 20 feet. Using ratios of all the bones in the wing, Frey found that the wings of pterosaurs became proportionally longer with respect to the size of the shoulder girdle as they grew.

When a Brazilian fossil collector discovered a fragment of ornithocheirid pterosaur shoulder girdle that was more than twice the size of the same bone in the 16-foot wingspan individual, it became clear that ornithocheirids achieved immense wingspans. Frey's calculations showed that this pterosaur must have had a wingspan of between 30 and 40 feet. The 30-foot figure is considered an underestimate, with 33 to 36 feet being most likely. Of course, this one example need not have been the largest...that specimen probably still remains to be discovered!

Iguanodon: perhaps the most successful dinosaur

While migrating across both North America and Europe, the *Ornithocheirus* in *Walking with Dinosaurs* flies over herds of *Iguanodon*, a two-ton herbivorous dinosaur that has been recorded from most of the Northern Hemisphere including North America, Europe, and Asia. In fact *Iguanodon* appears to have been the most abundant dinosaur of the Early Cretaceous and one of the most widespread and successful types from the whole of dinosaur history. It has been studied in great detail by Dr. David Norman of Cambridge University, and together with the abundance of specimens, this has made it one of the best-known of all dinosaurs. Dr. Norman has studied nearly all aspects of *Iguanodon*'s biology, including its anatomy, its feeding mechanism, the way it walked and ran, what it looked like and its distribution. He has also examined many closely related dinosaurs and the evolutionary history of the group. Apart from *Archaeopteryx*, perhaps no other dinosaur has received so much scientific attention.

How did *Iguanodon* walk?

In *Walking with Dinosaurs*, *Iguanodon* is shown walking slowly on all fours, but running at speed on only its hindlegs. Both alternatives

▷ **Perhaps the most successful dinosaur of them all, *Iguanodon* is known from Early Cretaceous rocks found throughout the Northern Hemisphere. The largest individuals of *Iguanodon bernissartensis*, the largest species, may have grown to 40 feet.**

seem likely given the anatomy of this dinosaur and the evidence from fossil tracks attributed to it. The vertebrae over the hip region of *Iguanodon* have elongate spines that are crisscrossed by overlapping bony tendons forming a reinforcing network that would have helped hold both the front of the body and the heavy tail in a horizontal position. The strength and size of the hindlimbs in this dinosaur indicates that *Iguanodon* was quite capable of walking bipedally. The tail acted as a counterbalance for the head and body when the animal walked this way.

△ Tracks show that, like many herbivorous dinosaurs, *Iguanodon* moved in herds. *Iguanodon* and its footprints are especially numerous in the rocks of southern England, Germany, and Spain.

Walking with Dinosaurs shows *Iguanodon* standing bipedally when fighting, feeding, or surveying its surroundings. Fossil footprints attributed to *Iguanodon* on the Isle of Wight, England, are mostly of the large three-toed kind made by the hindfeet but occasionally the crescent-shaped print of the hand is also preserved. In holding its body horizontally, it appears that *Iguanodon* could also use its stout and well-muscled forelimbs in supporting its weight, and therefore it was also able to walk quadrupedally.

The skin of *Iguanodon* is known from several specimens preserved as impressions made in fine-grained mud. These show that its skin was scaly like that of most reptiles.

What preyed on *Iguanodon*?

Like many living herbivores, *Iguanodon* surely fell victim to the aggressive predators of its time. *Walking with Dinosaurs* depicts a scene where an *Iguanodon* is attacked by the theropod *Utahraptor*. Evidence for this interaction is currently lacking, but we do know that *Iguanodon* did fall prey to several other types of theropod. In the famous dinosaur fossil beds of the Isle of Wight, Steve Hutt of the local museum found the remains of an *Iguanodon* associated with *Neovenator*, a theropod that resembled the *Allosaurus* from *Time of the Titans*. It is thought that this *Neovenator* died while scavenging on the *Iguanodon* carcass.

On further examination, Hutt noticed something quite remarkable about the *Iguanodon* bones. The neural spines of the vertebrae had large deformities in the form of big loops of bone and irregular, lumpy bony growths. Pathologist Dr. David Cooper believes that these are bones that had only partially healed after being bitten in two. The new bone growth shows that the animal survived the attack,

THE DISCOVERY OF *IGUANODON*

IGUANODON WAS THE first herbivorous dinosaur ever to be described, and thus it is all the more fitting that it should have featured in *Walking with Dinosaurs*. The first remains were described by the medical doctor Gideon Algernon Mantell of Lewes, Sussex, in 1825. At first only the teeth were known, and although he realized that these were most likely from a giant herbivorous reptile, a better idea of their identity did not arise until Mantell noted their superficial similarity with the teeth of modern iguanas. Subsequent finds of bones allowed Mantell to attempt a reconstruction, and he surmised that *Iguanodon* was an immense lizardlike quadrupedal animal. This reconstruction was immortalized by Benjamin Waterhouse Hawkins in the form of two concrete models displayed in the grounds of the Crystal Palace, London.

Not until the 1870s did many complete, articulated specimens of *Iguanodon* come to light, this time in a coal mine at Bernissart, Belgium. The Belgian zoologist Louis Dollo now concluded that *Iguanodon* was a bipedal giant reptile, built something like an immense kangaroo. Dollo's reconstruction held sway until late in the twentieth century. A more accurate view of *Iguanodon*'s appearance has arisen since the 1960s as experts have come to appreciate that dinosaurs had a predominantly horizontal body posture. In the 1980s Dr. Norman was able to show that *Iguanodon* was in fact well suited for some walking on all fours, and thus Mantell was, at least in part, right.

but it must surely have been in tremendous pain for a long time. It may also have been rather disabled, and perhaps this made it easy prey for the *Neovenator.*

Polacanthus, a spiky armor-plated herbivore

Another dinosaur that makes an appearance in *Walking with Dinosaurs* is the spiky, armour-plated *Polacanthus*, a member of a group of herbivorous dinosaurs called ankylosaurs. In *Giant of the Skies* we first meet this armored quadruped on the North American leg of the *Ornithocheirus* migration. However, the first specimens of *Polacanthus* were discovered on the Isle of Wight, England, in 1865. Some specimens discovered in the 1840s also appear to have been from *Polacanthus,* but were left in a Hackney carriage on their way to a museum!

So far only a few *Polacanthus* specimens have been discovered in any state of near completeness, and the skull is still poorly known even today. In contrast, the armor of *Polacanthus* is well preserved and shows that this dinosaur had triangular plates of armor along the length of its tail. As in the tail spikes of *Stegosaurus,* these appear to have projected sideways. Bands of armor covered the back and neck and large spikes projected upwards and sideways from among these bands. The upper surface of the hips was covered in a sheet of polygonal armor scutes that were fused together, forming a strong protective shield. Thus, like other ankylosaurs, *Polacanthus* was an impregnable dinosaurian armor-plated fortress.

Did *Polacanthus* live in North America?

Though *Walking with Dinosaurs* showed *Polacanthus* living in North America, definite evidence of this is lacking. However, very similar to *Polacanthus* is a dinosaur named *Gastonia,* found in Early Cretaceous rocks of Utah. *Gastonia,* described by Dr. James Kirkland and colleagues as recently as 1998, is known from much more complete remains than *Polacanthus,* and several well-preserved individuals, with complete skulls, are known. The skull of *Gastonia* is short, forming a broad triangle when seen from above. This new information allows us to predict that the skull of *Polacanthus* would be similar. Thus it can be argued that the head of *Polacanthus* in *Walking with Dinosaurs*

△ Well armored with its triangular spikes and hip shield, *Polacanthus* was probably a low-level browser, as suggested by its short neck and body shape.

is probably a little too long and narrow. New data frequently overturns palaeontological ideas, as is true of all sciences, and paleontologists often have to revise their reconstructions in the light of new information.

Another North American relative of *Polacanthus* is *Hoplitosaurus*. This dinosaur is so similar to *Polacanthus* that some scientists have suggested that they are the same, and that *Polacanthus* was common to both North America and Europe during the Early Cretaceous.

Utahraptor, vicious predator of the Early Cretaceous

▷ Known only from Utah, *Utahraptor* was a giant relative of the famous *Velociraptor*. It had huge, curved claws on both its hand and feet and a stiff tail for balance.

A third dinosaur in *Giant of the Skies* is the theropod *Utahraptor*. We meet this ferocious predator on the European side of the young, newly opening Atlantic Ocean. As its name suggests, *Utahraptor* was first discovered in Utah. It has not been discovered yet in Europe, but there is no reason why it should not occur there since other Early Cretaceous dinosaurs, like *Iguanodon* and perhaps *Polacanthus*, were common to the two regions.

Fossils of predatory dinosaurs are rare, and so far only three near-complete skeletons of theropods have been found in Early Cretaceous rocks of Europe. These include *Baryonyx*, famous for its elongated

skull and enlarged thumb claw, and *Neovenator*, both from England, and the multi-toothed *Pelecanimimus* from Spain. However, there are lots of fragmentary theropod remains such as isolated vertebrae and limb bones that could be from animals closely related to *Utahraptor*.

Utahraptor belongs to a group of theropods known as dromaeosaurs, sickle clawed predators with birdlike folding arms and enlarged brains. Most of these were small as dinosaurs go, being between three and ten feet in length. *Velociraptor* from Mongolia, star of the book and film *Jurassic Park*, was one of the smaller members of this group (although in *Jurassic Park* the animal reconstructed was actually based on *Deinonychus*, a ten-foot-long relative from the Early Cretaceous of North America). *Utahraptor* was one of the exceptions, being a giant perhaps 16 feet long and about half a ton in weight. Other giant dromaeosaurs are known from the Cretaceous of Japan and Mongolia. While smaller forms, like *Deinonychus*, may have been formidable predators that could dispatch *Iguanodon* sized dinosaurs with ease, *Utahraptor* was exceptional and could theoretically have tackled giant prey including the immense sauropods.

Dromaeosaurs were rather lightly built and had large eyes, tails stiffened by overlapping rodlike bony projections of the vertebrae,and a very large raised claw on the second digit of the foot. This claw was more than twice as long as the others on the foot. Unlike many other theropods, such as tyrannosaurs and ornithomimids, dromaeosaurs had short and stocky legs with extra muscle attachment sites on the thigh bone. These legs may have given them a powerful burst of speed and would also have provided extra strength for the slashing kicks we suppose were required in disembowelling their prey.

What was *Utahraptor's* raised foot claw used for?

The second toe of *Utahraptor* is shown in *Walking with Dinosaurs* as being held raised up off the ground during walking. This was initially assumed because there is evidence that dinosaur claws were covered by a keratin sheath, as they are in all living animals. Indeed, *Rahonavis*, a dromaeosaur or primitive bird from the Late Cretaceous of Madagascar, still preserves traces of this keratin sheath on its sickle claw. Because keratin is easily worn down, the claw must have been raised off the ground in order to keep the tip sharp. The anatomy of the dromaeosaur foot shows that the second toe was specially built to keep the claw in this retracted posture, and fully articulated specimens now

DROMAEOSAURS IN EUROPE?

WALKING WITH DINOSAURS showed the large dromaeosaur *Utahraptor* living in Spain. However, this dinosaur's fossils are only known from North America. Were dinosaurs related to *Utahraptor* present in Europe at this time? This seems likely, given that *Iguanodon* and *Polacanthus*-like ankylosaurs were common to both areas, but it has proved hard to verify.

While examining some remains from the Early Cretaceous of the Isle of Wight, England, supposed to belong to pterosaurs, paleontologists Stafford Howse and Dr. Andrew Milner discovered that one specimen, named *Ornithodesmus cluniculus*, had been misidentified as a pterosaur but belonged instead to a small meat-eating theropod dinosaur. Detailed comparisons of this specimen with other theropods suggest that it might represent a dromaeosaur, though this would require the discovery of more complete remains before verification is possible.

Dromaeosaurs are now known from the Late Cretaceous of Europe, although the remains are highly fragmentary compared with some near complete North American specimens. In 1998, Dr. Jean Le Loeuff and Dr. Eric Buffetaut announced the discovery of a European dromaeosaur which they called *Variraptor*, after the region in southern France where it was found. This demonstrated unequivocally that dromaeosaurs were present in Europe, though their presence in the Early Cretaceous still remains to be confirmed.

confirm that this was indeed the case. Of course, this would mean that dromaeosaur footprints would only leave the impressions of two toes. Two-toed footprints have now been reported from Asia and are thought to have been made by dromaeosaurs.

Walking with Dinosaurs also shows *Utahraptor* using its second-toe claw as a weapon when preying on *Iguanodon*. The shape and size of this claw strongly suggest that it was used for this function. However, direct evidence that dromaeosaurs used the sickle claw as a weapon comes from paleopathology, the study of ancient injuries and diseases. Researcher Cynthia Marshall and colleagues recently reported a dromaeosaur toe bone that was clearly fractured during life, and had later healed. This injury, sustained to what is considered to have been the main offensive weapon of this group of dinosaurs, supports an idea proposed earlier by Professor John Ostrom that dromaeosaurs used this claw in attacks on other dinosaurs.

Utahraptor was only described in 1993 when Dr. James Kirkland and colleagues announced it as the largest of the dromaeosaurs. The large sickle claw on the foot of *Utahraptor*, when reconstructed with its keratinous sheath, measures 11 inches around its outer curve. This is about five times the size of the largest tiger's claws – such a massive claw on this powerful, agile predator must have been a formidable weapon. Unless *Utahraptor* hunted in packs, the death of an *Iguanodon* at the hands of a *Utahraptor* would have been a bloody and painfully prolonged affair.

How did *Utahraptor* use its hands?

Utahraptor in *Walking with Dinosaurs* is also seen using its hands when hunting, and folding its hands backwards when they are not in use. These reconstructed movements result from studies on the bones, joints, and possible ranges of movement in the hands, wrists, and arms of other dromaeosaurs (not enough is known from the forelimbs of *Utahraptor* for it to be studied in this way). All dromaeosaurs have elongated arms with three-fingered hands and mobile wrists. A special half-moon shaped wrist bone would have allowed them to fold their hands up against the side of the body, in the same way that birds fold their wings. These and other strong similarities indicate that dromaeosaurs are the closest known relatives of birds, and perhaps their ancestors.

Were dromaeosaurs flightless birds?

Though many paleontologists regard dromaeosaurs as possible bird ancestors, others point to problems with this idea. In some ways dromaeosaurs are more like modern birds than is the Late Jurassic *Archaeopteryx*, usually thought to be the first bird. For example, dromaeosaurs have fully back-turned hips like birds, unlike *Archaeopteryx*, and also a huge breastbone, again in contrast to *Archaeopteryx* (the breastbone of *Archaeopteryx* is very small and is only known from one of the seven specimens). These features led dinosaur expert Gregory Paul to suggest in 1984 that dromaeosaurs were actually flightless birds that had descended from an *Archaeopteryx*-like ancestor.

Although birds first make their appearance in the Late Jurassic, the Early Cretaceous saw the appearance of the first forms that resembled the birds of today. Early Cretaceous birds have been found in China, England, Spain, and even in the same deposits in Brazil where ornithocheirid pterosaurs have been found. Most Early Cretaceous birds were small, not much larger than today's thrushes, and often smaller. Most belong to a group called the enantiornithines, a word meaning "opposite birds," so named because some of their bones fuse together in a direction opposite to that seen in modern birds.

Early bird from the age of the dinosaurs

The bird that appears in *Giant of the Skies* is *Iberomesornis* from the Early Cretaceous of Spain. The Spanish fossil deposits have yielded several different birds, showing that even in the Early Cretaceous birds were diverse, though their fossils are rare. *Iberomesornis* was small, with a total length of around three inches. Unfortunately its skull has not yet been discovered, so it could have been a little longer. Like modern birds it had a pygostyle, a series of fused vertebrae at the tip of its tail,

▽ ***Archaeopteryx*, preserved with all of its feathers, was an animal intermediate between theropods like *Utahraptor* and modern birds.**

◁ *Iberomesornis* was an early "opposite bird" from Spain. It still had clawed fingers. Whether these early birds nested in trees or on the ground remains unknown.

and so was much closer to modern birds than to the Jurassic *Archaeopteryx.* No feathers were found associated with *Iberomesornis* when it was discovered, but other birds from the same deposits have been found with feathers and many isolated feathers have also been discovered there.

Unlike modern birds, but like *Utahraptor* and other coelurosaur theropods, *Iberomesornis* still had three, clawed fingers growing from its wing. Cretaceous birds that are preserved with their feathers in place show that feathers grew from the upper surface of the second finger, but that the first and third fingers were still free. Relatives of *Iberomesornis* further show that the first finger now supported feathers and was destined to become the alula, a mobile structure used by birds to control the flow of air over their wing surface. While an alula is known for some enantiornithines, we do not know if it was present in *Iberomesornis.*

FEATHERS

IBEROMESORNIS FROM *WALKING with Dinosaurs* is one of the earliest birds and is reconstructed with feathers because distinctive skeletal features show that it belongs to the same group of animals that includes *Archaeopteryx*, from the Late Jurassic of Germany, and living birds. Thanks to the very fine-grained rocks in which it is preserved, we know that *Archaeopteryx* has true feathers identical in microscopic structure to those of living birds. But what are feathers, and how are they used?

Feathers are highly elaborate structures embedded in the skin of birds that provide insulation, protection from sharp objects, color for mate attraction, camouflage, nesting material, and (in birds that fly) provide streamlining, and the necessary lightweight surface for flying. They are light and tough. Although they are shed on a regular basis, with careful maintenance involving preening and parasite prevention, feathers can last for more than one year. Not bad considering they are dead structures composed of a type of keratin.

A problem for biologists has been explaining the evolution of feathers. They are only known today for birds, and the closest living relatives of birds, the crocodiles, certainly do not possess feathers, nor any structure that could be considered a protofeather. As it is now clear that birds evolved from dinosaurs, and in particular from dromaeosaurs or close relatives, a question arises: did dinosaurs also possess feathers?

The oldest feathers found in the fossil record are those of *Archaeopteryx*. Although *Archaeopteryx* is a bird, it is also a respectable dinosaur. It has jaws with teeth, a pelvic girdle much more similar to that of theropod dinosaurs than to that of a modern bird, and it has an elongated tail composed of numerous small vertebrae. Modern birds on the other hand lack teeth, have a pelvis highly modified so that all of its components project backwards and are fused together, and have a tail that has been reduced to a little nub of bone called the pygostyle.

The similarity *Archaeopteryx* has with some theropods might therefore suggest that feathers appeared first on dinosaurs and were inherited from them by birds. So, did all dinosaurs possess feathers? Did feathers on non-flying dinosaurs look like feathers on birds? There are many questions that still remain to be answered, but discoveries of small predatory dinosaurs in Liaoning Province, China, have recently shown that at least some dinosaurs were truly feathered.

The Early Cretaceous world: setting the stage for today

Giant of the Skies focused on the last migration of one Early Cretaceous animal, the giant pterosaur *Ornithocheirus*. Though *Ornithocheirus* itself is an amazing creature, its journey served largely to illustrate the increasingly complex interactions that were appearing between plants and animals in this ancient world.

Early Cretaceous fossils include some of the best studied of all Mesozoic fossils. For the herbivorous dinosaur *Iguanodon*, in particular, we can be more confident about its appearance, lifestyle, and anatomy than we can about virtually any other dinosaur.

Walking with Dinosaurs showed that, while pterosaurs were very much the dominant animals of the Early Cretaceous skies, birds were now flourishing and appeared destined for future success.

Ornithocheirus and its relatives were large, sometimes enormous, gliding pterosaurs that hunted fishes and other sea creatures. Some forms possessed crests on their beaks and controversy continues over whether or not these were indicators of sexual dimorphism. Argument also continues among paleontologists over the function of these intriguing structures. Did they have specific functions – such as use as rudders, sails, or keels during flight – or were they only for display? Given the size of the head crest in some species of *Tapejara*, the last idea might seem unlikely, but think of some of the display structures seen in certain living animals. Male peacocks put their lives at risk by having enormous, heavy tail feathers that mean that they are not as effective at escaping from predators as other pheasants. Did *Tapejara* and other crested pterosaurs risk their lives in similar ways?

Exceptionally well preserved Early Cretaceous rocks, such as those of Liaoning Province in China, not only preserve birds and theropods with feathers, they also reveal diverse assemblages of plants, insects, and other animals. In fact, many essentially modern forms of life were becoming firmly established in Early Cretaceous times. Flowering plants had evolved, and numerous kinds of insect now specialized in feeding on them. Fossil insects from Russia and Australia show that fleas and other parasites were present, and sucking the blood of mammals and other animals of the time. The rest of the Cretaceous would see increasing complexity in these communities.

CHAPTER 5

Spirits of the Ice Forest

106 MILLION YEARS AGO

During the time of the dinosaurs, the Earth's climate was moderate overall. Over most of the planet there were no sharply defined hot and cold seasons and there were no ice caps at the North and South Poles. The Early Cretaceous was a largely tropical time in which dinosaurs thrived on every land mass. Though the southern continent of the Jurassic, Gondwana, had started to break up, Australia and New Zealand were still connected to Antarctica and were far south of their present location. Fossil sites in southern Victoria, Australia, have yielded abundant Early Cretaceous dinosaurs and other animals and form the focus of the *Spirits of the Ice Forest* episode.

Small bipedal dinosaurs lived here, along with small mammals and giant aquatic amphibians. These animals were featured in *Walking with Dinosaurs*, as were a variety of other animals that might perhaps have come to the area seasonally. These include large predatory dinosaurs, the large and unusual herbivorous dinosaur *Muttaburrasaurus*, and large pterosaurs. Relatively little is known of these animals, which is unfortunate because they might have exhibited some unusual features and behaviors that could have been the result of their adaptation to life in this southern climate. What is known, and how this was applied to *Walking with Dinosaurs*, is addressed here.

How cold was the Cretaceous South Pole?

During the Early Cretaceous, Victoria was located at about 77° south. Despite being so close to the South Pole, this area did not suffer the incredibly low temperatures that occur at this latitude today. Not only was the Mesozoic world warmer overall, but the circumpolar ocean that keeps Antarctica cold today did not exist. The edge of the Antarctic continent was located just over the South Pole, thus the waters of the polar seas were open to the warm waters of the Pacific Ocean.

◁ Though only known from a handful of specimens, it is possible that *Muttaburrasaurus* was a herding dinosaur. Next to nothing is known about the biology of this animal.

Studies of different forms of oxygen (isotopes) in the Cretaceous sandstone deposits of southern Australia suggest temperatures ranging from between 21°F and 41°F (-6°C and +5°C). This is cold enough for frosts and perhaps the presence of glaciers on high ground, though some geologists contest the evidence that it got this cold.

Other scientists have used the plants and animals in the fossil assemblages as rough guides to temperature. Dr. Robert Spicer and

Dr. Judith Parrish examined the structure of some of the Cretaceous southern plants and concluded that the mean annual temperature was more likely around 50°F (10°C). Furthermore, lungfish were present here at this time, and living members of this group cannot breed in water colder than 50°F (10°C). However, it might be dangerous to assume that fossil plants and animals had the same limits of temperature endurance as their living relatives. Recently, it has been shown that some of the ground at these southern sites was frozen as it is in seasonally icy high latitudes today, supporting the evidence for the lower temperature estimates and suggesting that winter temperatures might have dropped to as low as -22°F(-30°C).

Regardless of the controversy over temperature, there is no debating that these polar areas would still have had between three and five months of nearly total darkness, as there is at the poles today. The fossil plants known from the area show adaptation to cold or dry conditions (the dryness perhaps brought about by the freezing of the water during winter) and either have specially thickened outer cuticles, or were deciduous and would have dropped their leaves at the onset of the dark season. Animals living in this harsh environment would have adapted one of two strategies. They could either have been well adapted to survive in the cold and dark, or they would have had to migrate elsewhere for the winter.

Dinosaur Cove and the animals of *Walking with Dinosaurs*

Thanks largely to the research of a team led by two Australian paleontologists, Dr Thomas Rich and Dr. Patricia Vickers-Rich, a startling array of Cretaceous polar animals have been discovered by more than two decades worth of excavation on Victoria's southeast coast. It is a site called Dinosaur Cove, ironically given this name two years before any dinosaur bones were found there, that has provided the most amazing insight into Cretaceous life near the South Pole. Dinosaur Cove and nearby sites have yielded a variety of bipedal, plant-eating ornithopod dinosaurs, a handful of predatory dinosaurs, large aquatic amphibians, and many other animal fossils. Several of these animals were reconstructed for *Walking with Dinosaurs.*

It might seem odd that these animals flourished in an apparently cold, seasonally dark climate. However, it would also have been perpetually light during the warm season, allowing rampant plant growth. The continual summer daylight would also have allowed more time for foraging, feeding, and breeding, so, as on the northern tundras today, it would have been advantageous for animals to make the most of this opportunity while it lasted. How they survived the dark times remains largely unknown.

The Early Cretaceous plant fossil record shows that this area was forested. Podocarps, ginkgoes, and members of the monkey puzzle group were the dominant trees and are thought to have created a fairly open canopy structure. Cycads, horsetails, ferns and related plants dominated nearer the ground – flowering plants were not important here, though they would be later on in the Cretaceous. These forests, with their diverse plants and animals, therefore created

◁ How cold the winters were in the forests of the Cretaceous South Pole remains controversial. Some evidence suggests low, freezing temperatures and the presence of snow and ice.

▷ If the temperature really did plummet to well below zero in the polar winter, the dinosaurs that remained here would have had to cope with snow and ice.

an ecosystem type that does not exist today – i.e. temperate woodlands subjected to seasonal darkness.

Fossil insects and spiders are well represented in the rocks around Dinosaur Cove. Bugs, beetles, and flies were abundant and dragonflies, crickets, and wasps are also known. In *Walking with Dinosaurs*, a weta – a kind of large, flightless cricket from New Zealand – was shown as living in the polar forests. It might be that the ancestors of these herbivorous insects lived across Antarctica, southern Australia, and New Zealand in the Cretaceous, and that they only became restricted to New Zealand once it became isolated at the end of the Cretaceous.

Many fossil turtles have been found in the region, where they had lived alongside a variety of bony fish. Such aquatic animals would have been prey for larger aquatic animals, including the small plesiosaurs known from Dinosaur Cove, and the giant amphibians. Birds were present in the area, but are known only from their feathers.

Mammals of the Cretaceous South Pole

In *Walking with Dinosaurs*, a small mammal attempts to raid a dinosaur's nest. Mammals are unknown from Dinosaur Cove but for a single tooth. Their fossils are, however, better known from another famous Australian site of similar age, Lightning Ridge in New South Wales. One of these Lightning Ridge mammals, *Steropodon*, is an early relative of the duckbilled platypus. As mammals from the age of dinosaurs go, *Steropodon* was very large, being roughly comparable to a modern badger. The only similarly sized Mesozoic mammal was *Didelphodon* from the Late Cretaceous of North America.

The second Lightning Ridge mammal, *Kollikodon*, is unique, but is also a member of the group of mammals (called monotremes) that include the platypus and the spiny termite-eating echidnas. *Kollikodon* had broad, flattened teeth, each with four rounded cusps – indeed, when first discovered it was nicknamed "Hotcrossbunodon"! It appears to have eaten objects that required crushing before they could be swallowed, such as snails or hard seeds.

A third Early Cretaceous mammal was *Ausktribosphenos*, this time from Victoria and, though not from Dinosaur Cove, from a nearby locality. *Ausktribosphenos* was tiny, about mouse sized, and equipped with tiny pointed teeth that appear suited for a diet of insects and other small animals. Despite its small size, *Ausktribosphenos* has created a storm of debate. Dr. Rich and his colleagues, the team who first described this animal in 1997, suggested that it was the earliest known member of the group of mammals called placentals – the group to which we, and all other living mammals except monotremes and marsupials (kangaroos and their relatives), belong. The proposal that placentals first appeared in Australia is very controversial, as all other fossil evidence indicates that this is a Northern Hemisphere group that did not appear in Australia until long after the Cretaceous. Other experts on early mammals have therefore argued that *Ausktribosphenos* is really a monotreme, or a member of a group of primitive Mesozoic mammals called symmetrodonts. Debate over the identity of *Ausktribosphenos* continues – a reminder of how theories about evolution and the fossil record may always be challenged by new discoveries.

Leaellynasaura: diminutive dinosaur from the polar forests

Small ornithopod dinosaurs were among the first fossils from Dinosaur Cove to be studied and have since become the best-known animals from the site. Consequently, one of these dinosaurs – *Leaellynasaura* – was reconstructed for *Walking with Dinosaurs.*

Rich and Vickers-Rich found that there were several forms of small ornithopod at Dinosaur Cove, and that they could be distinguished by differences in their teeth. As in other ornithopods, the leaf-shaped teeth of these dinosaurs show that they were plant eaters. The toothless tips of their lower jaws indicate the presence of a beak used to cut foliage. There was a corresponding beaked area at the tip of the upper jaw, though in small ornithopods there were still teeth in this part of the jaw.

The tooth and jaw differences seen among the small Australian ornithopods suggest that they were eating different kinds of plants, or chewing in different ways. *Leaellynasaura* was the first of these ornithopods to be named, and was named by Rich and Vickers-Rich after their daughter, Leaellyn, who had participated in its discovery. Thousands of *Leaellynasaura* bones are now known from Dinosaur Cove and many of them appear to be from juveniles. This suggests that *Leaellynasaura* bred and raised its young in the polar forests, presumably during the productive summer.

Leaellynasaura's large eyes

Leaellynasaura appears to have been a small dinosaur, between six and ten feet in length, which was remarkable in having particularly huge eyes. This implied that it had acute eyesight, and perhaps that it was adapted for seeing in low light levels. A cast of the brain of *Leaellynasaura* shows that the brain was large and had well-developed optic lobes, the parts of the brain devoted to sight.

However, some of the first specimens found do not seem to have been fully grown as several of their teeth have not yet erupted, and the bones in their skulls do not seem to have been fully fused together. Isolated thigh bones of *Leaellynasaura* show that the especially big eyed specimens are small and only about one-third grown. One characteristic feature of baby and juvenile animals is that they have proportionally huge eye sockets. If the big eyed specimens are juveniles, therefore, it seems likely that adults might not have had such enormous eyes. Also, study of other small ornithopods shows that they have rather similarly proportioned eyes to *Leaellynasaura*, so this Australian dinosaur might not have been so exceptional after all.

How do we know that *Leaellynasaura* lived in groups?

Leaellynasaura is depicted in *Walking with Dinosaurs* as living in complex social groups. Obviously we will never know the full complexities of any dinosaur society, but various pieces of evidence have allowed scientists to make inferences about the lifestyles of these animals. Evidence that small ornithopods like *Leaellynasaura* traveled and perhaps lived in groups comes mainly from two sources. First, multiple fossil tracks that appear to have been made by small *Leaellynasaura*-like dinosaurs suggest that the animals were social. The tracks show several individuals moving together in the same direction.

Secondly, the fossils of small ornithopods are often found together. In the Early Cretaceous rocks of the Isle of Wight, the similar small ornithopod *Hypsilophodon* is known from groups of individuals that appear to have died together, perhaps in a flood or when they became mired in quicksand. Similarly, the Late Jurassic form *Drinker* has been found in groups of individuals that apparently died together when

◁ *Leaellynasaura* is abundant at the Victorian site Dinosaur Cove and may well have been social, living and moving in groups.

▽ The skull bones of *Leaellynasaura*, seen here from above, show that this dinosaur had large eyes and a brain with large optic lobes.

the burrows they were hiding in became flooded. Although *Drinker* is the only small ornithopod known to have definitely used burrows, it is possible that *Leaellynasaura* did the same. This could explain how *Leaellynasaura* survived the polar winter.

One small bipedal ornithopod discovered at Dinosaur Cove (it is not possible to determine whether or not it is the same as *Leaellynasaura*) has a badly broken leg, yet this has clearly healed and the animal seems to have survived subsequently for some length of time. It seems amazing that a bipedal dinosaur that is supposed to rely on speed to escape from enemies survived with such an injury for so long. Perhaps this is further evidence that these dinosaurs lived in social groups and helped look after one another.

What would *Leaellynasaura* have done in the winter time?

One of the mysteries of the small Dinosaur Cove ornithopods is how they survived the cold, dark winter. *Walking with Dinosaurs* depicted the animals as taking refuge in the forest, rather than migrating. Living small animals the size of these dinosaurs generally do not migrate as it is too expensive in terms of their energy requirements, though short-distance migrations are not such a problem. One suggestion is that *Leaellynasaura* moved nearer to the edge of the Antarctic Circle during the cold, dark season, where the climate was possibly more amenable. However, because of the far southern location of the Dinosaur Cove area, even this would have required a monumental round trip of 500 miles or more. An alternative idea would have been simply to migrate the much shorter distance to more coastal areas where the maritime climate would have prevented the winters from becoming too cold. Larger dinosaurs, like the muttaburrasaurs that also feature in *Walking with Dinosaurs*, would not have been faced with this energy dilemma and could well have migrated greater distances.

If *Leaellynasaura* did stay in the deep south, would it have hibernated during the cold season? Rich, Vickers-Rich, and colleagues have recently presented evidence that indicates that *Leaellynasaura* did not hibernate, but stayed active throughout the southern winter season. By taking cross sections of *Leaellynasaura* bones, they could examine what happened during the animal's growth, and it was clear that the animals did not undergo any periods when they stopped growing. Such periods create distinctive circles, called lines of arrested growth,

FURRY LITTLE DINOSAURS?

ONE INTERESTING QUESTION that arises is that if these polar dinosaurs were living in a place of extreme cold, might they have been covered in fur or feathers to help them stay warm? Exceptionally well preserved fossils of small predatory theropod dinosaurs from Early Cretaceous rocks of Liaoning Province, China, show that these animals were covered either in true feathers, or in bristlelike structures that would have resembled fur. Skin impressions from large theropods and large ornithopods like *Iguanodon* show that these dinosaurs had scaly skin.

However, exceptional fossils preserving the skin impressions of small ornithopods are as yet unknown – with one possible exception. One skin impression from a small ornithopod was described by the famous paleontologist Charles Gilmore in 1915. Unfortunately, this specimen has since been lost, but Gilmore noted that it had a "punctured" surface. One expert has suggested that this "puncturing" might indicate that the animal's skin sported some kind of furry or feathery covering.

Regardless of whether *Leaellynasaura* and its relatives were feathery, furry, or scaly, it might also be likely that they were insulated from the cold by body fat. We may never know from fossils, but perhaps *Leaellynasaura* grew plump during the productive summer and then used its body fat for warmth and as a food reserve during the cold polar winter.

in the internal fabric of bones, a little like the growth rings seen in tree trunks. These circles are seen in animals that hibernate and in slow-growing animals that stop growing during times of hardship. Their absence in *Leaellynasaura* means that we can be fairly confident that *Leaellynasaura* was growing, and therefore active, all year round.

If *Leaellynasaura* was remaining active year-round in this seasonally cold climate, perhaps it could generate its own body heat. The bones of other ornithopod dinosaurs are well supplied with blood vessels and do not show lines of arrested growth either. Furthermore, they show that the animals grew quickly. Perhaps these features indicate that *Leaellynasaura* and all of its relatives were "warm blooded".

Predatory dinosaurs of the polar forest

Predatory dinosaurs, the theropods, existed in the polar forests of the Cretaceous and are represented in *Walking with Dinosaurs* by a large predator of an undetermined species. Like all the Cretaceous polar theropods, this species is poorly known and represented only by scrappy fragments of bone. A second form is represented only by a shin bone that appears similar to a Northern Hemisphere group of theropods called the ornithomimids, a group often compared to living ostriches because of their long necks and legs and toothless, ostrich-like heads. Rich and Vickers-Rich named this possible Australian

▽ On the edge of the lush polar forest, a theropod finds a *Muttaburrasaurus* carcass. Like all living predators, predatory dinosaurs would have scavenged on occasion.

ornithomimid *Timimus* after their son, Tim! A third Dinosaur Cove theropod is known only from a piece of jaw and some vertebrae. Again, these fragments are surprising, as they most resemble the corresponding bones of another Northern Hemisphere group, the oviraptorids. Oviraptorids had toothless, somewhat parrotlike skulls and might have been omnivores or leaf-eaters.

The large theropod, the one featured in *Walking with Dinosaurs*, is best known from a broken astragalus, the main ankle bone in dinosaurs and the one that, via an upward pointing triangular flange, was locked firmly into position against the front of the animal's shin bone. Exactly what kind of theropod this astragalus comes from has been the subject of much dispute.

Was the large polar theropod an allosaur?

When first described in 1981 by Dr Ralph Molnar, Dr. Timothy Flannery and Thomas Rich, this bone was thought to be very similar to that of *Allosaurus*, the Late Jurassic predator of North America and Portugal. The find was therefore interpreted as proof of the presence of *Allosaurus* in the Early Cretaceous of Australia. There were some differences from the astragalus of the known species of *Allosaurus*, suggesting that this was a different species, and it seemed

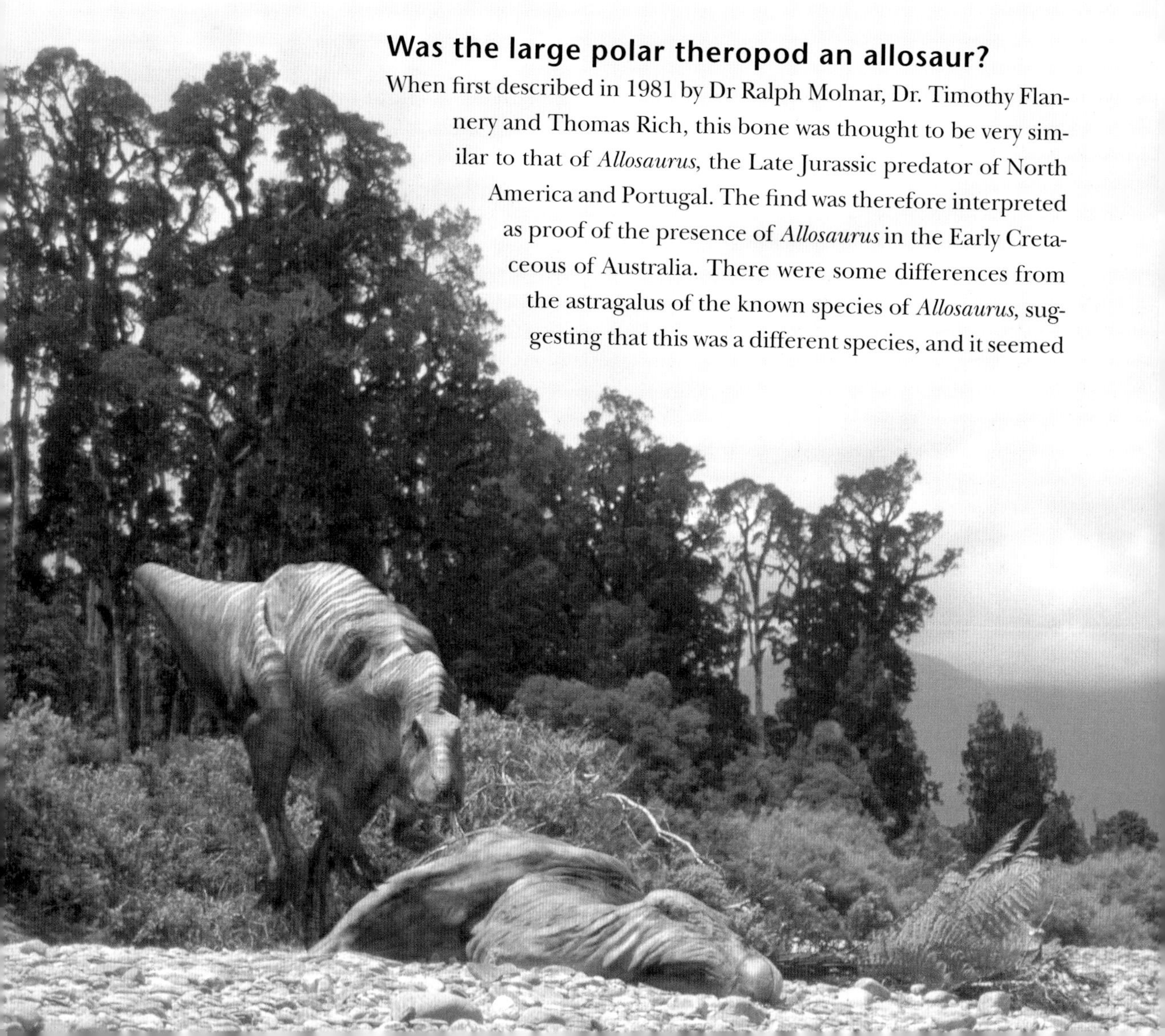

◁ Scrappy bits and pieces show that theropods were present in the polar forests, but exactly what kind of theropods remains controversial. The idea that allosaurs may have been present here is the subject of argument.

to have come from a smaller animal than the Northern Hemisphere allosaurs. Perhaps, the team suggested, *Allosaurus* had survived as a dwarf form in the forests of the Southern Hemisphere while it had become extinct elsewhere.

Dr. Samuel Welles, a noted expert on the anatomy of theropod bones – and in particular their ankles! – challenged this identification in 1983. Welles listed nineteen differences between the Australian specimen and the astragalus of *Allosaurus*, and concluded that it could not be from *Allosaurus*, and might instead belong to a relative of the ostrichlike ornithomimids. Molnar, Flannery, and Rich responded to Welles' comments in 1985. They contested the supposed differences pointed out by Welles and argued again that the bone seemed so like that of *Allosaurus* that it most probably was from a Southern Hemisphere representative of this genus. An analysis of the bone by Dr. John Long in 1998 resulted in the same conclusion – Long found that its proportions were more like those of *Allosaurus* than of any other theropod dinosaur.

The very latest development in this controversy comes from Daniel Chure, a world authority on the anatomy of *Allosaurus*. As part of a global scale investigation into *Allosaurus* and all of its relatives, Chure examined the Australian bone. He concluded that the specimen was not from *Allosaurus* after all, and that it could not even be shown to be from a theropod closely related to *Allosaurus*. Most other experts now accept his view. The bone does seem to be from a large, presumably predatory theropod, and one that was part of the same group of advanced theropods as *Allosaurus*. However, it does not appear to be from *Allosaurus*.

Koolasuchus – killer newt from hell

While the Australian theropod may not, after all, be a "living fossil" of its time, other animals of the deep south certainly were. Another main character of the *Walking with Dinosaurs* polar forest episode is *Koolasuchus*, an immense predatory amphibian belonging to a group called temnospondyls.

Temnospondyls were huge aquatic amphibians, something like enormous salamanders but often with crocodile-like heads. Though many kinds were only three feet long or less, the largest were monsters of 20 feet and more. These animals had been diverse and abundant in the waterways of the Permian and Triassic periods. Indeed, in the environments in which dinosaurs first appeared, such as the Chinle Formation in the southwestern USA, temnospondyls were still among the dominant predators of the aquatic realm. In the rocks of the latest Triassic, however, temnospondyls are absent, and were thus thought to have become extinct.

A discovery in Australian Jurassic rocks in 1940 changed all that: *Austropelor*, known only from its lower jaw, was discovered in Queensland. To begin with, *Austropelor* caused a controversy. Some paleontologists argued that it had been misidentified, and was not really a temnospondyl, while some geologists thought that the rocks from which it had come had been wrongly dated, and must in fact have been of Triassic age. Still others suggested that the fossil had been eroded from Triassic rocks, and then entombed later in younger, Jurassic rocks (a geological process called reworking). However, more discoveries of temnospondyls in these rocks, and a thorough geological investigation of their dating showed that the original identification was correct. A second Jurassic temnospondyl from Queensland, *Siderops*, was named and described in 1983.

These surprising finds were followed by other Jurassic temnospondyls from China and Russia. What these animals were doing surviving in places, and at a time, when they had been thought long extinct, seemed a mystery. Contrary to expectations, the Jurassic temnospondyls could also be shown to be unrelated to one another – they actually represented three different groups that had all, independently, survived across the Triassic boundary. Greater surprises were yet to come, as in 1981 Dr. Flannery and Dr. Rich announced the discovery of a temnospondyl lower jaw in the Early Cretaceous rocks of Victoria, Australia. This large specimen had been originally identified as the lower jaw of an armored dinosaur!

Later discoveries of vertebrae and skull bones confirmed that the jaw was from a temnospondyl. One of the vertebrae showed the unmistakable and distinctive shape common to temnospondyls, while the skull bone showed the distinctive ornamented surface texture typical of this group. Enough was now known from this Australian amphibian for it to be given a name – *Koolasuchus* – an appellation that commemorates the discovery of some of these bones by Lesley Kool of Monash University, Victoria.

The probable lifestyle of *Koolasuchus*

Though not as big as some of the earlier temnospondyls, *Koolasuchus* was a formidable giant that would have been around 16 feet in length. It belongs to a group of short-headed temnospondyls that were especially well adapted for almost perpetual life in water. Some of these appear to have retained the external gills they possessed during their larval, or tadpole, stage into adulthood. Retention of larval features into adulthood, called neotony, is seen today in some salamanders, that, like *Koolasuchus* spend all of their life in water. Their skull bones reveal shallow canals that, like identical canals on the skulls of modern fishes, must have been connected to nerves and used in detecting the movement of water currents. By sensing such water motion, temnospondyls would quite literally have felt the nearby movements of fishes and other prey animals.

Koolasuchus and its relatives also had their eyes on the top of their rather flattened heads, located so that they pointed upwards. The same configuration is seen in the living clawed frogs. They sit on the mud at the bottom of pools, and owing to their upward-pointing eyes can see prey animals as they swim above. When prey does appear, they rise towards the surface and quickly open their large mouths, the sudden in-rush of water helping to capture the prey by a form of suction. *Koolasuchus* was certainly not fast enough to pursue fast-moving fishes and other prey, so it is thought to have hunted in the same manner.

How did *Koolasuchus* swim?

Koolasuchus and related temnospondyls all have a generally similar overall shape. Their skulls are huge, blunt-snouted, and broad, being somewhat broader than the width of the body. Their mouths were correspondingly broad and their jaws lined with numerous small, conical teeth. These appear well suited for grasping slippery prey like fish. In contrast to their huge mouths, the limbs of these temnospondyls were small and very weak, implying that they could not move on land, or only with great difficulty and for a very short distance. Complete temnospondyl skeletons also reveal a deep, flexible tail that would have provided the main thrust during swimming. Their spines show that their bodies would have undulated from side to side when swimming and perhaps when walking, if they were able to do this.

SECRETS OF THE LOST TEMNOSPONDYLS

WHY *KOOLASUCHUS* SURVIVED in the cold south, and how it managed to do so, has been an area of some speculation. One possible explanation is that temnospondyls might have competed with crocodiles, and that the evolution of river crocodiles in the Jurassic pushed the temnospondyls from their realm except in the deep south, a place supposedly too cold for crocodiles to thrive.

This idea is problematic since nearly all Jurassic and Early Cretaceous river-dwelling crocodiles were small, and it seems unlikely that they would have competed with huge temnospondyls. Also, fossil crocodiles are now known from the southern environments. However, temnospondyls would have started out life as tiny larvae. At this time they would have been vulnerable to predators, so maybe early crocodiles and other predators do account for the disappearance of the temnospondyls.

Another idea concerns the environment in which the last temnospondyls lived. Writing about the discovery of Jurassic temnospondyls in Russia, the Soviet paleontologist Dr. Alexandrovich Nessov noted that, here, the temnospondyls lived in an estuarine environment and were found alongside fishes and other aquatic animals that also seemed particularly closely related to Triassic forms. Estuarine environments pose special physiological problems for aquatic animals as the changing salinity of the water means that the animals have to be able to cope with the extremes of salty and of fresh water. If temnospondyls and other animals had become adapted to this environment in the Triassic, other animals would then have had a hard time (in evolutionary terms) replacing them.

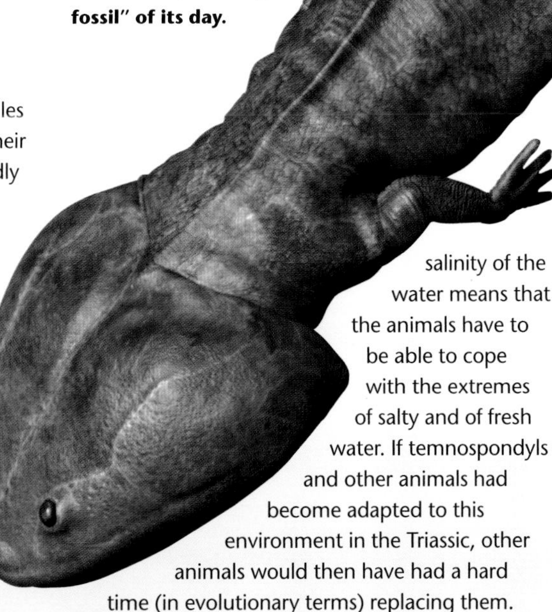

▷ **The giant *Koolasuchus*, a "living fossil" of its day.**

△ **With upward-facing eyes and a system of sensitive skull canals, *Koolasuchus* was probably an ambush hunter that sat at the bottom of pools, waiting for prey to come within range.**

Though these fossil finds show that temnospondyls had survived for longer than first thought, they were still rare and restricted to only a few environments. Perhaps they could only survive in "safe haven" places where key competitors, parasites or predators were unable to survive. In agreement with this, work on fossil marine invertebrates like starfishes has shown that, because there are fewer predators in polar regions, they serve as havens where species made extinct elsewhere can survive.

The survival of giant temnospondyls into the Jurassic and Cretaceous does suggest that they were far more hardy than first thought. Perhaps the fact that they survived for so long was due to an ability to tolerate a range of climatic conditions, to go for long periods without food, or to go into torpor (an inactive state similar to hibernation) during times of hardship. Modern amphibians, which are descended from temnospondyls according to some experts, have a similar ability to survive harsh conditions.

Though temnospondyls did survive for a very long time, they are not known from the fossil record of the Late Cretaceous.

Rare pterosaurs of the deep south

Flying reptiles, pterosaurs, were present in this Cretaceous southern land mass and are shown flying over the polar forest in *Walking with Dinosaurs*. These polar forms could prove to be particularly interesting as they might have been specially adapted for life in this unusual seasonal environment. Unfortunately, these pterosaurs are very rare as fossils, so our knowledge of them is poor. A single leg bone is known from the Victorian deposits, and bones that appear similar to *Ornithocheirus* are known from the Early Cretaceous of Queensland. This indicates a form with a 6.5 foot wingspan. The front of the lower jaw of this Queensland pterosaur is known and shows an elongated, toothed snout much like that of *Ornithocheirus*.

A pterosaur shoulder girdle, perhaps from the same individual, was found about 550 yards away at the same site. This has similarities to the shoulder girdle of *Ornithocheirus*, but it is also much like that of the Late Cretaceous *Pteranodon*. If the specimen does represent a relative of *Pteranodon*, it is one of the earliest members of the group and, if the toothed jaw belongs with it, might show that early members of this group were equipped with large conical teeth. *Pteranodon* is completely toothless, but the existence of *Pteranodon*-like pterosaurs with teeth has recently been confirmed by a discovery in Cretaceous rocks in Brazil. A complete pterosaur skull, representing a new kind, as yet unnamed, has both the bony head crest of *Pteranodon* and the elongated teeth of *Ornithocheirus* and its relatives.

Another Queensland specimen, described in 1987, is a partial pelvis. This might also belong to an *Ornithocheirus*-like form, but again it has some similarities with a *Pteranodon* pelvis. Unlike nearly all other pterosaur pelvic girdles, it is uncrushed and allows us to see the correct orientation of the hip socket. This is important as it shows the exact disposition of the thigh bone, which has been much debated.

Further Southern Hemisphere pterosaurs are known from New Zealand, but this time from the Late Cretaceous. Represented only by a partial wing bone and a tooth, it is difficult to be certain of their identity. Nevertheless they are important in being the most southerly record yet known for pterosaurs, and in showing that Cretaceous pterosaurs were living in a cool, seasonal environment. Again, this has led to suggestions that these animals must have been "warm blooded" if they were to survive in such climates.

PTEROSAUR HIP BONES

AS IS TRUE OF some other especially well preserved pterosaur hip girdles, the specimen discovered in Queensland has a hip socket directed sideways. This shows that the thigh bone was also directed sideways. This posture means that this kind of pterosaur did not have its legs directly underneath its body and, therefore, was not built for bipedal running. Dinosaurs and birds, which clearly could, and can, run bipedally, have erect thigh bones that point downwards from their hip sockets and are positioned alongside their hips during normal locomotion.

The horizontal inclination of the thigh bone in pterosaurs also implies that the leg was connected to the main wing membrane, either at the knee or the ankle. This is confirmed by pterosaurs where the wing membranes are preserved. This information allowed the pterosaurs in *Walking with Dinosaurs* to be reconstructed with extensive wing membranes that were connected to the hind legs. *Walking with Dinosaurs* also shows the pterosaurs walking on all fours and not, as some paleontologists had suggested, walking on just their hind legs.

Having the wing membranes connected to the legs would not mean that pterosaurs were clumsy or inefficient in the air. This is proven by bats, many of which are faster and more agile in the air than certain birds. Neither would they have struggled on the ground. Again, many bats are highly adept walkers despite their wing membranes. Abundant fossil footprints thought to have been left by pterosaurs also suggest that they were quite able on the ground, though clearly not as agile or fast moving as dinosaurs and birds.

Muttaburrasaurus – strange name, strange dinosaur

The large and enigmatic dinosaur *Muttaburrasaurus* was a huge herbivore superficially like *Iguanodon*. Named after Muttaburra in central Queensland where the first specimen was discovered, *Muttaburrasaurus* had a large, hollow bump taking up much of its snout. What was *Muttaburrasaurus* doing with this nose bump? It seems reasonable to speculate that it used this structure to make resonating calls. *Muttaburrasaurus* also had strongly built, three-toed feet and slim arms that could have been used to help support its weight. Early reconstructions provided *Muttaburrasaurus* with a thumb spike, as there is in *Iguanodon*. However, the single, poorly preserved, broken structure found with *Muttaburrasaurus*, and originally interpreted as a thumb spike, is hardly convincing. Furthermore, new evidence shows that *Muttaburrasaurus* is not closely related to *Iguanodon*. Dr. Ralph Molnar, one of the original describers of *Muttaburrasaurus*, now argues that it lacked a thumb spike.

Muttaburrasaurus is not a well-known dinosaur. Only three specimens have so far been discovered, one of which is only a skull. Some isolated teeth thought to belong to *Muttaburrasaurus* are known from Lightning Ridge in New South Wales. What is particularly interesting is that the two skull specimens are notably different. In the skull of the first specimen, the bump on the snout rises fairly gently from just ahead of the eyes. The second specimen is from slightly older rocks and has a taller bump that arises more abruptly in front of the eyes. The teeth of the two specimens are also different. Perhaps these differences show that there were different species, or perhaps they represent the two sexes.

The exact affinities of *Muttaburrasaurus* are not known. When first named and described in 1981, it was thought to belong to the same group of plant-eating dinosaurs as *Iguanodon*, a family called the iguanodontids. Consequently, many artistic reconstructions produced at this time depict *Muttaburrasaurus* as, effectively, an Australian *Iguanodon*. Since then, however, it has become clear that *Muttaburrasaurus* is very different from the iguanodontids – it is, in fact, strikingly unique.

▽ ***Muttaburrasaurus* had a peculiar nose bump, perhaps used in improving the resonance of its calls or in visual display.**

△ Some track sites show mixed herds of herbivores that were perhaps attracted to the same food source. Here, the large muttaburrasaurs eat from the higher foliage while the leaellynasaurs forage on the ground.

What did *Muttaburrasaurus* eat?

In *Walking with Dinosaurs Muttaburrasaurus* is shown eating tough vegetation and breaking branches in its quest for plant food. Unfortunately, not enough is known of this dinosaur to be at all confident about what it really ate, but its skull bones do provide some clues. Unlike *Iguanodon* and its relatives, *Muttaburrasaurus* had teeth where the main cutting edge is restricted to the outside margins. This indicates that it used its teeth to shear up food by vertical slicing. This method of chewing is also seen in the horned dinosaurs, but as there is no doubt that these are not closely related to *Muttaburrasaurus*, this represents a case of convergent evolution – that is, where unrelated animals evolve similar solutions to a problem. *Muttaburrasaurus* is also unusual in that the bony bar behind its eye, the postorbital bar, is remarkably broad. This implies that it was using its postorbital bar to resist stresses placed upon the skull. Exactly what these stresses might have resulted from, however, is unknown – perhaps from slicing especially tough branches during feeding.

A reanalysis of the *Muttaburrasaurus* skeleton in 1996 led Molnar to argue that, not only was *Muttaburrasaurus* unrelated to the *Iguanodon* family, it also seemed outside the group that includes the *Iguanodon* family and all of its closest relatives. It seems instead to represent part of a group that split from other ornithopods in the middle of the Jurassic, probably shortly after the ancestors of *Leaellynasaura* first evolved. In the absence of true *Iguanodon*-like ornithopods, perhaps *Muttaburrasaurus* was able to become superficially like one and to play the same role that iguanodontids did elsewhere in the world.

One final observation is Molnar's discovery that *Muttaburrasaurus* shares some unusual features in its lower jaw and teeth with *Atlascopcosaurus*, one of the small bipedal ornithopods from Dinosaur Cove. This raises the intriguing idea that both belong to a special Australian family of ornithopod dinosaurs.

Known as it is from only a handful of individuals, we cannot say anything definite about the lifestyle and habits of *Muttaburrasaurus*. In *Walking with Dinosaurs* it is portrayed migrating in a small herd. Perhaps its apparent presence in both Queensland and New South Wales indicates that it was migratory, but this can only be inferred.

The fate of the Cretaceous polar dinosaurs

Geological data and analysis of magnetic characteristics in the rocks show that, by the end of the Early Cretaceous, Australia had broken away from Antarctica and was moving northwards. Indeed, it continues to do so at the rate of a few inches a year, so in theory Australia will collide with southeast Asia in the distant geological future. Though this northward movement would have meant an end to the low polar temperatures of the Cretaceous forests, it would also have ended the long, productive light season and would have resulted in the disappearance of this unique ecosystem. Dinosaurs probably continued to thrive in Australia until the end of the Cretaceous, but the unique polar forms like *Leaellynasaura* became extinct at this point.

Presumably, the same kinds of dinosaur as those found in southern Australia also occurred in what is now Antarctica, and maybe they could have survived there for longer. Antarctica was not to become truly ice covered until well after the end of the age of dinosaurs, and Late Cretaceous discoveries show that dinosaurs including ornithopods and armored dinosaurs were still living there at this time. New Zealand, which broke away from the side of Antarctica at the end of the Cretaceous, was also inhabited by dinosaurs until the very end of that period. It was not a gradual climate change that put a final end to their diversity, but rather it was events of a different kind.

Final word on the polar forests: what we know, and what we would like to know

As we have seen, the polar forests of the Cretaceous represented a unique and strange environment, and one that was an irresistible, though challenging, subject to focus on for *Walking with Dinosaurs.* The discovery of a rich variety of animals, large and small, allowed the reconstruction of a complex community where interactions can be inferred. The discovery of abundant *Leaellynasaura* specimens shows that these small herbivores were common, and that they exploited the rich plant growth. Large predatory dinosaurs are known, and quite likely preyed on *Leaellynasaura* and other species. *Koolasuchus* was an amazing surprise when first discovered, and was presumably an imposing freshwater predator in the rivers of the time.

However, in discussing what is known about these remarkable creatures, we are often as not discussing what we would like to know, but do not. Thanks to the discovery, preparation, and examination of the many *Leaellynasaura* bones from Dinosaur Cove, inferences can be made about its growth rate, seasonal behavior, and, perhaps, how well suited it had been for polar darkness – as determined by features of its skull bones. This is actually a lot of information in comparison to what is known for most other fossils of the time. For example, the life cycle of *Koolasuchus* is totally unknown and, while we can make guesses based on fossils of distant relatives, is open to speculation. Known *Koolasuchus* material is also incomplete, but nearly complete skeletons of other members of the same group do show how this animal looked and moved. The pterosaurs of this area are also poorly known, as are the theropod dinosaurs.

In the absence of much key information, *Walking with Dinosaurs* brought this environment and its animals to life by making some educated guesses. The end result is a believable, complex community where animals had evolved to exploit a seasonally productive, but seasonally hostile, environment.

CHAPTER 6 Death of a Dynasty

65 MILLION YEARS AGO

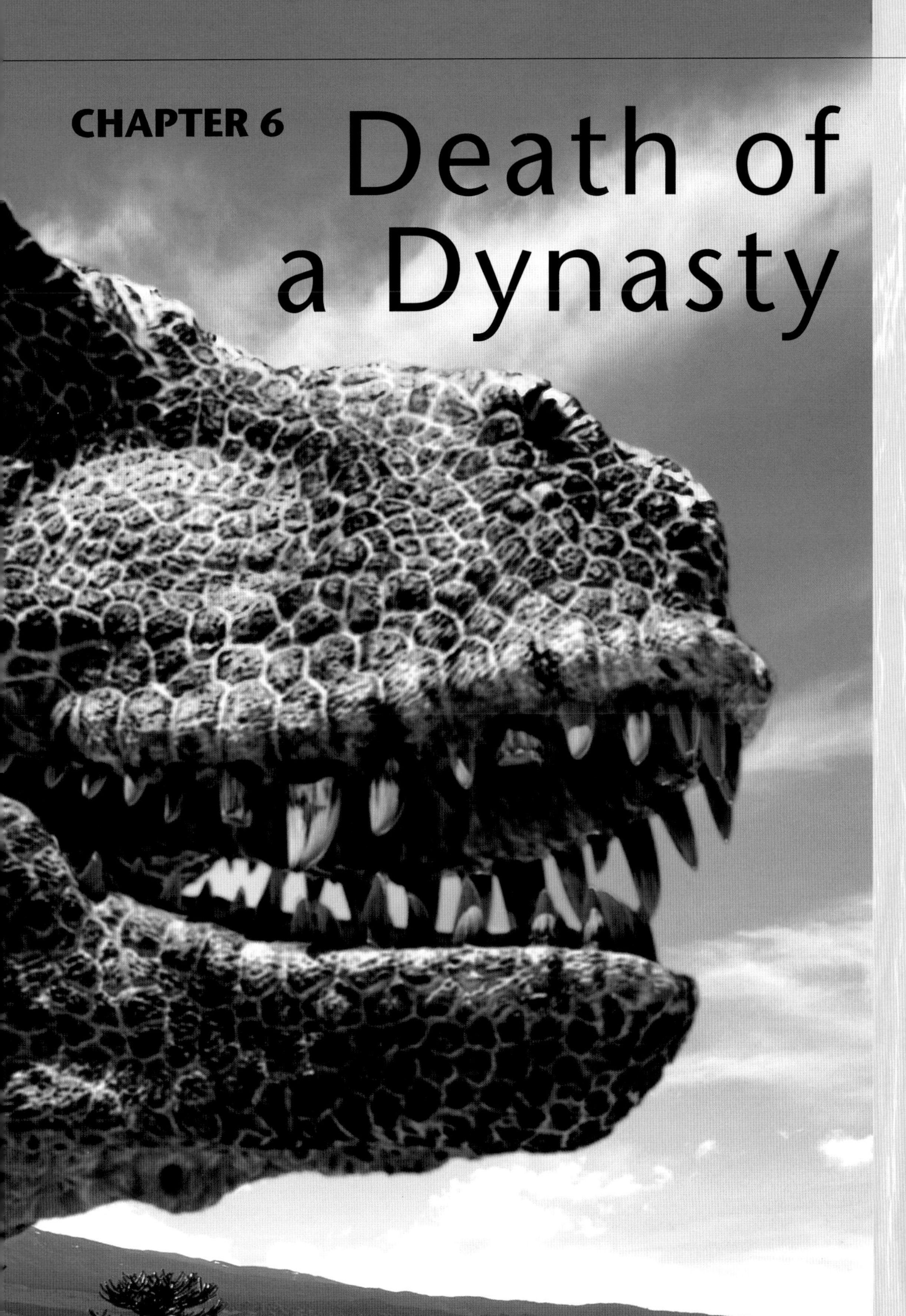

DEATH OF A DYNASTY, *Walking with Dinosaurs*'s final episode, is set at the very end of the Cretaceous, shortly before the entire Age of Dinosaurs was to come to an end. Thanks to spectacular discoveries made largely in the western half of North America, the dinosaurs of this time have become the best-known of the whole of dinosaur history. *Walking with Dinosaurs* focuses in particular on one Late Cretaceous animal community: that of the Hell Creek Formation of Wyoming, Montana and the adjacent parts of North America. This is where *Tyrannosaurus rex*, one of the biggest land-living predators of all time and the main character in *Death of a Dynasty*, was first discovered. *Tyrannosaurus* lived alongside, and preyed on, the giant horned dinosaurs *Triceratops* and *Torosaurus*, and we also meet these in *Death of a Dynasty*.

Duckbilled herbivorous dinosaurs called hadrosaurs, descendants of animals similar to *Iguanodon*, were also abundant in the Hell Creek and were also preyed upon by *Tyrannosaurus*. The peculiar long-headed hadrosaur *Anatotitan* – its name means "giant duck" – features in *Death of a Dynasty*. Giant armor-plated dinosaurs, related to *Polacanthus* from *Giant of the Skies*, also appear in the program, and we focus on the giant *Ankylosaurus*, famous for its stiffened, club-tipped tail. Smaller dinosaurs lived alongside these Hell Creek giants. *Dromaeosaurus*, a swift long-armed theropod much like *Utahraptor* from *Giant of the Skies*, is also shown in *Death of a Dynasty*.

Throughout the Age of Dinosaurs, pterosaurs, the flying reptiles, had flown over the heads of the dinosaurs. During the Late Cretaceous, however, this group had declined in diversity and, at the time in which *Death of a Dynasty* is set, only a handful of forms existed. These kinds were remarkable, however, in being huge: they were almost certainly bigger than any pterosaurs from earlier times, their wingspans exceeding even those of the ornithocheirids from earlier in the Cretaceous. In *Death of a Dynasty* we see the best-known and geologically youngest of these giants: *Quetzalcoatlus*. We also see the giant alligator *Deinosuchus*.

◁ The massive, deep head of *Tyrannosaurus* – the most formidable land predator of the Late Cretaceous. Its massive jaws, stout, and curving teeth provided it with an awesome biting force.

Smaller kinds of animal were flourishing in the Late Cretaceous. Small mammals, some of which were related to living animals like marsupials, were fairly diverse in the Hell Creek and were doing many of the things that small mammals do today – some burrowed underground, some climbed in trees and others scurried around in the undergrowth. One of the largest Late Cretaceous mammals,

Didelphodon, stars in *Walking with Dinosaurs* as an opportunistic predator and scavenger that, whenever possible, raids dinosaur nests and eats from carcasses. An important group of reptiles related to lizards, the snakes, first appear in the Late Cretaceous and perhaps owe their success to the increasing diversity of small mammals. Though the giant dinosaurs were destined to become extinct at the end of the Cretaceous, birds – which had evolved from small theropod dinosaurs in the Jurassic period – were becoming ever more abundant.

Diverse and complex as it was, this Late Cretaceous world was to come to an abrupt end. An awesome cataclysm was to bring the final closure of the Age of the Dinosaurs.

The world in Late Cretaceous times

Death of a Dynasty is set at a time in which dinosaurs had been the dominant large animals on land for 160 million years. They had lived from pole to pole and had filled all environments (except those of lakes, rivers, and the sea) from deserts, forests, and plains to lake shores, marshes, and the cold woodlands of the polar regions. Though small dinosaurs, including diminutive theropods and plant-eating ornithopods similar to *Leaellynasaura*, were still present in the Late Cretaceous, several groups had now evolved fantastic giant forms. Were we around to see them, these dinosaurs might have seemed like the culmination of their groups' evolutionary histories.

Whatever happened to the sauropods?

In *Time of the Titans*, we saw how the world at the end of the Jurassic period had been dominated by the long-necked sauropods. Where were they now, in a world apparently ruled by tyrannosaurs and horned dinosaurs? Had they become extinct? Sauropods are all but absent from the fossil record of Late Cretaceous North America, and as a consequence, it was thought for a long time that they had declined virtually to extinction following their Jurassic heyday. However, investigations of the Late Cretaceous fossil records of Europe, South America, and Africa have shown that sauropods were in fact abundant and diverse right up to the end of the Cretaceous. Some of these Late Cretaceous sauropods were actually as big as or bigger than the largest Jurassic forms.

Late Cretaceous North America therefore appears to have been an exceptionally strange continent, not an ordinary one, and its animal community was certainly not reflective of animal communities elsewhere in the world. Indeed, the horned dinosaurs, tyrannosaurs, and duckbilled hadrosaurs that dominated Late Cretaceous North America were largely unique to this land mass, though some of these dinosaurs were shared with eastern Asia.

Introducing *Tyrannosaurus*

Death of a Dynasty's main character is *Tyrannosaurus*, the most famous and perhaps the most awesome predator of all time. Reaching over 40 feet in length and more than five tons in weight, *Tyrannosaurus* was a formidable monster specially adapted to kill other giant dinosaurs. *Walking with Dinosaurs* paints a complex picture of this dinosaur's life. Its sensory and killing abilities are reconstructed as it hunts and searches for a mate, we see its nesting behaviour, and how females cared for, fed, and protected their babies. A range of studies on the biology, mechanics, and lifestyle of *Tyrannosaurus* allowed *Walking with Dinosaurs* to reconstruct these aspects of the animal's life.

There is no doubt that *Tyrannosaurus* was the dominant predator in the Hell Creek community. It was first named by the American paleontologist Henry Fairfield Osborn in 1905, who immediately recognized its publicity potential. It was quite clearly the largest predatory dinosaur – and largest predatory land animal – then known. Osborn's name for the species, "tyrant lizard king," has become the best-known and most memorable dinosaur name of them all.

▷ Growing to more than twelve metres and five tons, *Tyrannosaurus* was one of the biggest theropods of all time. Some African and South American theropods might have been slightly bigger. Unlike these forms, *Tyrannosaurus* had long, slim legs.

In contrast to those theropods that were lightly built and long-necked, *Tyrannosaurus* was a massive, thickset beast with a huge, barrel-shaped body, a strongly muscled thick neck, and an enormous, deep skull. One spectacularly well-preserved *Tyrannosaurus* skeleton, the most complete *Tyrannosaurus* specimen presently known, has yielded a wealth of information. This specimen is "Sue," a huge individual from South Dakota that has been part of a legal battle between the Sioux Indians, on whose land Sue was found, and the paleontological team who collected the fossil, led by expert collector Peter Larson.

Studies on the chevron bones beneath Sue's tail have led to the

conclusion that she is a female (see page 30). This suggests that female tyrannosaurs were larger and more heavily built than males. Sue is also important in exhibiting a catalogue of injuries including broken tail bones and ribs, a damaged face and upper arm, and a fractured shin. Many of the injuries appear to have been inflicted by other tyrannosaurs, perhaps in territorial disputes or during mating. Some of these injuries had healed, but others might have been responsible for Sue's death. Indeed, one *Tyrannosaurus* specimen, also from South Dakota, seems to have been eaten by another one. This might be evidence of opportunistic cannibalism – one *Tyrannosaurus* just happened to find the carcass of another – or it could show that these animals frequently killed one another, as do some predators today.

Recent studies show that many tyrannosaur skulls have bites made by other tyrannosaurs. One individual even has a tooth embedded in the outside of its lower jaw, suggesting that tyrannosaurs fought by standing face-to-face and biting at one another's jaws. Life as a tyrannosaur was tough.

The life appearance of *Tyrannosaurus*

Walking with Dinosaurs accurately portrays the remarkable body shape of *Tyrannosaurus.* Despite its size and weight, its legs are proportionally elongated and slim. Its tail is proportionally short compared with earlier large theropods like *Allosaurus* and its body is also short and fairly stiff. All tyrannosaurs had remarkably short arms compared with most other theropods, and *Tyrannosaurus* was one of the shortest-armed members of the tyrannosaur group.

The small size of the forelimbs in *Tyrannosaurus* has led some paleontologists to suggest that they were all but useless, and on their way to being lost altogether. Perhaps the immense skull was so heavy, and so devastating a weapon, that the forelimbs were now obsolete and simply added extra weight to the front of the animal's body. New studies, however, show that the arms were well muscled and very powerful. It also appears that the two fingers of the tyrannosaur hand were nearly opposable, and would have functioned something like tongs. Despite these newly revealed details, scientists remain unsure about the function of these specialized arms.

In *Walking with Dinosaurs, Tyrannosaurus* is provided with a complex head that has rounded hornlets over the eyes and a ridge across the cheek region. This reflects the fact that the skull of *Tyrannosaurus*

HOW FAST COULD *TYRANNOSAURUS* MOVE?

IN *WALKING WITH Dinosaurs*, *Tyrannosaurus* is shown as a heavy animal that moves with a real sense of weight to its gait. It is also mostly shown walking at a moderate speed. Both of these deductions are based on the large size of the animal – it seems likely that it was not particularly sprightly. However, *Walking with Dinosaurs* also shows *Tyrannosaurus* racing in for the kill when chasing an *Anatotitan*. Here it is swift and fleet-footed. Evidence that tyrannosaurs could move fast, at least on occasion, comes from their long, slim legs with their massive muscle attachment sites, and their ankle and foot bones. These are specially locked together for strength and suggest that at least some fast movement was possible.

Footprints that appear to have been made by *Tyrannosaurus* were discovered in New Mexico in 1993. They do not show the animal moving at great speed, but the likelihood of actually having bursts of speed preserved in the fossil record is relatively slim – nearly all animals spend the vast majority of their time moving around at their normal walking pace. Exactly how fast *Tyrannosaurus* would have moved when running at full speed is actually an area of contention. Some paleontologists think that the long, well-muscled legs and locked-together foot bones show that *Tyrannosaurus* was a sprinter that could run as fast as modern ostriches or horses, whereas other have argued that its leg bones were not strong enough for anything faster than a gentle trot. A 1995 study calculated that, if a *Tyrannosaurus* were to stumble and trip while sprinting at speed, the fall would have been fatal.

Though there is no way of being sure, most experts today do not think that *Tyrannosaurus* was an ostrichlike sprinter, or a slow plodder, but something in between. It was quite probably capable of bursts of speed during which it would have exceeded the top speed of a running human.

is not simply a rectangular box, as often shown in old reconstructions, but a complex shape with hornlets in front of and behind the eyes and a low, bumpy ridge along the top of the snout. Some experts think that the hornlets in front of and behind the eyes were united to form one large horn – hence the structure seen in *Walking with Dinosaurs* – but others argue that this is incorrect.

The senses of *Tyrannosaurus*

Walking with Dinosaurs shows *Tyrannosaurus* clearly reacting to visual stimuli – that is, things that it can see. For example, we see the nesting female chase off small mammals and dinosaurs. The back of the skull in *Tyrannosaurus* is much broader than the muzzle, and the eye sockets are directed forwards. These features suggest that, unlike most other large theropods, *Tyrannosaurus* was equipped with a visual field where, just in front of its snout, both of its eyes perceived the same area. This is also the case in humans and is called binocular vision. Because both eyes see the same area, but from different angles, binocular vision allows animals to be especially good at judging distances. *Tyrannosaurus* was presumably, therefore, adept at delivering precision strikes with its awesome mouth. The spaces for the eyeballs show that its eyes would have been huge, and casts of its brain show that, as seems true of most dinosaurs, the optic lobes were well developed. The eyesight of this giant was therefore acute.

Walking with Dinosaurs also shows *Tyrannosaurus* using other senses. The female is shown calling when in search of a mate – therefore, acute hearing was inferred to be an important sense in *Tyrannosaurus.* The tyrannosaurs in *Walking with Dinosaurs* are also shown sniffing out carrion, so a well developed sense of smell is also depicted.

New studies on the skull of "Sue" by Dr. Chris Brochu provide the evidence behind these inferred sensory capabilities. Sue reveals a delicate stapes – the bone that conducts sounds in the ears of reptiles – showing that the hearing of *Tyrannosaurus* was well-developed and acute. This suggests that it might have used vocalizations to communicate with other members of its species. There is therefore some justification for the long distance calls the female *Tyrannosaurus* makes in *Walking with Dinosaurs.* Brochu's studies also show that the parts of the brain devoted to smell in *Tyrannosaurus* were huge – several times wider than the main part of the brain (the cerebrum) in fact. The sense of smell in this dinosaur was, therefore, exceptionally well developed and *Tyrannosaurus* could almost certainly have located carrion from a distance, and perhaps used scent as part of its communication repertoire.

△ The skull of "Stan", a *Tyrannosaurus* from South Dakota. The enlarged bosses in front of and behind the eyes are evident; the teeth are partially out of their sockets and are thus unrealistically long.

How did *Tyrannosaurus* kill and eat?

Tyrannosaurus in *Walking with Dinosaurs* is an awesome predator that kills its prey by biting at the body or neck – killing by inflicting massive tissue damage with its tremendously powerful jaws and stout teeth. That *Tyrannosaurus* could do this is now clear from fossil discoveries as well as studies on its teeth and skull bones.

First of all, evidence that *Tyrannosaurus* preyed on dinosaurs like *Triceratops* and *Anatotitan* is clear from tooth marked bones that have now been discovered. One hadrosaur skeleton provides clear evidence that, in life, it was attacked by a *Tyrannosaurus*. In this specimen, part of the top of the tail has been bitten by a *Tyrannosaurus*, leaving a bite-shaped wound. Somehow, however, the hadrosaur escaped, since the tail bones later healed. This is evidence that *Tyrannosaurus* fed on live hadrosaurs, and did not just scavenge on dead ones.

One find in particular inspired the *Tyrannosaurus* feeding behavior depicted in *Walking with Dinosaurs* and resulted in the re-enactment of a special feeding style. In *Walking with Dinosaurs* we see the female *Tyrannosaurus* feeding from a *Triceratops* that her mate has killed. While feeding, the female bites into the hip region and pulls flesh and bones away from the side of the carcass. This is called a "puncture and pull" feeding strategy. One *Triceratops* specimen in particular, a broken pelvis from Montana, has provided the best evidence that *Tyrannosaurus* fed in this way.

This pelvis is covered in tooth marks – nearly 80 in total – and large portions of the edge of the pelvis have been broken away. Putty casts made of the tooth marks prove that they were definitely made by *Tyrannosaurus* teeth. The deepest of the tooth marks penetrate the bone more than 4.5 inches, showing that *Tyrannosaurus* had a phenomenal ability to pierce bone. As revealed in 1996 by Dr. Gregory Erickson and Kenneth Olson, those tooth marks nearest the edges of the broken regions were associated with long score marks. Erickson and Olson concluded that the tyrannosaur had bitten into the bone, fastened on, and then pulled backwards with its body weight. The

◁ A baby *Tyrannosaurus* eats meat given to it by its mother. The association of baby *Tyrannosaurus* specimens with adults implies parental care, but this area is speculative.

pulling had created the score marks and had allowed the tyrannosaur to rip off those parts of the pelvis that are now missing.

Because the resistance of bone to the impact of hard objects is a known constant that can be measured, Erickson and colleagues were able to use these tooth marks to estimate the bite strength of *Tyrannosaurus.* In order to push teeth as deep into bone as had happened with the *Triceratops* pelvis, an amazing bite force of more than 13,000 newtons was required. By comparison, a man might have a bite force of about 700 newtons and a lion just over 4000. Only alligators, the animals with the strongest measured bite, have a force comparable with that estimated for *Tyrannosaurus.* These tests show that *Tyrannosaurus* had one of the strongest bites of any animal.

Bone crunching behavior in *Tyrannosaurus* is confirmed by the study of fossil dung or coprolites. A *Tyrannosaurus* coprolite was discovered in Late Cretaceous rocks of Saskatchewan in 1997. The large size of the coprolite, and the fact that it contained abundant bone fragments and was therefore produced by a carnivore, indicate that it was produced by a *Tyrannosaurus.* Also, no other giant carnivore is known to have lived in this place during the Late Cretaceous. The abundance of bone fragments in the specimen suggests that *Tyrannosaurus* was both capable of crunching up bones into small pieces, and in the habit of swallowing the pulverized fragments.

Was *Tyrannosaurus* a caring parent?

In *Walking with Dinosaurs,* the female *Tyrannosaurus* is shown as a caring parent that guards its eggs while they incubate in a warm mound of rotting vegetation. The female later protects and feeds its babies. Fossil nests are known for small theropod dinosaurs and a few examples even have adults preserved on top, still arranged in a brooding posture. These dinosaurs seem to have died when overwhelmed by sandstorms, or perhaps when smothered by mudflows.

No such outstanding evidence is available for large theropods like *Tyrannosaurus,* so what is shown in *Walking with Dinosaurs* is based on what is known about the nesting behavior of other dinosaurs. Presumably dinosaurs of *Tyrannosaurus'* size did not sit on their eggs – they would surely have crushed them – so perhaps they heaped up rotting vegetation on top of the eggs instead. The heat created by the decomposing plant material would have incubated the eggs, as it does in crocodile and scrub fowl nest mounds today. We do not know if

Tyrannosaurus actually guarded the nest but this seems likely given that all today's living relatives of dinosaurs (crocodiles and birds) do.

As for the matter of looking after the babies once they had hatched, as is also shown in *Walking with Dinosaurs*, some evidence is available though, admittedly, how it is interpreted is the subject of debate. Several adult *Tyrannosaurus* specimens, including "Sue", are preserved alongside juvenile *Tyrannosaurus* specimens. These instances might be proof of active parental care in tyrannosaurs (as appears to have been the case in *Allosaurus* – see page 65). One of the juveniles preserved near to "Sue" was well-grown, not a hatchling, and it might indicate that juvenile tyrannosaurs stayed with their parents for an extended period of time. On the other hand, it is equally likely that juvenile tyrannosaurs were fully independent and that the association of these juveniles with "Sue" is a coincidence. We simply need more evidence!

Dromaeosaurus, a small Late Cretaceous predator

Tyrannosaurus was the largest of the Late Cretaceous theropods, but another, smaller kind features alongside it in *Death of a Dynasty*. This was *Dromaeosaurus*, a dinosaur about the same size as a labrador with long arms with three-fingered hands, a stiffened tail and, like *Utahraptor* from *Giant of the Skies*, a raised sickle-shaped claw on the second toe of its foot (see pages 112-14). Like *Utahraptor*, *Dromaeosaurus* was almost certainly a dangerous predator and in *Walking with Dinosaurs* it is shown preying on small ornithopod dinosaurs and juvenile horned dinosaurs. *Dromaeosaurus* is not actually from the Hell Creek Formation but from an older Late Cretaceous rock unit called the Judith River Formation, but teeth and other fragments show that nearly identical forms did live in the Hell Creek community.

Unfortunately, the skeleton of *Dromaeosaurus* is all but unknown. Some leg and foot bones thought to be from *Dromaeosaurus* have now been re-identified as having belonged to a different, but related, kind of theropod. *Utahraptor*, which is known from a fairly substantial amount of skeletal material, is thought by some to be particularly closely related to *Dromaeosaurus*. Like other dromaeosaurs, *Utahrap-*

tor has stout, well-muscled legs, a raised sickle claw on the second toe of its foot, and three-fingered hands with enlarged and strongly curved hand claws. *Dromaeosaurus* was probably similar in appearance to this.

New discoveries of small theropods in Lower Cretaceous rocks of China show that all small theropods had featherlike structures covering their bodies. In light of this, the *Dromaeosaurus* in *Walking with Dinosaurs* should have been depicted with quill-like structures such as those seen on the *Ornitholestes* in *Time of the Titans*, rather than with the naked, scaly skin it was given.

Was *Dromaeosaurus* a social predator?

Walking with Dinosaurs showed two dromaeosaurs working together to split up a *Torosaurus* herd so that the vulnerable juveniles could be separated from their parents. This is obviously conjectural, but some evidence does suggest that dromaeosaurs cooperated in killing large dinosaurs. This evidence does not pertain to *Dromaeosaurus*, but to the Early Cretaceous dromaeosaur *Deinonychus*.

While uncovering *Deinonychus* material in the 1960s, Professor John Ostrom found that several of these dinosaurs were associated with the skeleton of the large herbivorous dinosaur *Tenontosaurus*. The *Tenontosaurus* skeleton was also associated with teeth that seemed to have been shed by several individuals of *Deinonychus*, presumably when they were feeding. Ostrom wondered how the dromaeosaurs had killed the tenontosaur when it was several times larger than they were. His conclusion was that *Deinonychus* worked together in packs and had bought down the *Tenontosaurus* together. However, Ostrom suggested that several of the pack members had been killed during the fight. Geological investigation has shown that these bones were not washed together by a river or flood.

While it is by no means impossible that *Deinonychus* and other dromaeosaurs did cooperate and hunt like this, other possibilities exist. Perhaps the dromaeosaurs exhibited mobbing behavior – that is, they did not live together permanently (like truly social animals, such as wolves and lions) but simply cooperated when prey was available. Some predatory lizards, crocodiles, and birds still do this today. True pack behavior for *Deinonychus* seems unlikely if it means that they routinely attacked an animal that usually ended up killing several of the pack members!

Ankylosaurus: a Cretaceous armoured tank

Besides the predatory dinosaurs of the Late Cretaceous world, *Walking with Dinosaurs* focused on the diverse dinosaur herbivores of the same time. With predatory theropods as dangerous as *Tyrannosaurus* and *Dromaeosaurus* around, these herbivores had to be either fast on their feet, well able to detect danger before it approached, or armored and capable of defending themselves. The ankylosaurs, represented in *Walking with Dinosaurs* by the "living tank"*Ankylosaurus*, clearly adopted this last option and developed it to the extreme. *Ankylosaurus* not only had large scutes, spikes, and plates covering the top of its head, back, flanks, and tail, it also had triangular horns projecting from the back of its head, a huge club on the end of its tail, and even mobile bony shutters within its eyelids!

▽ **The extensive body armor of ankylosaurs, like this *Ankylosaurus*, would have made them invulnerable to predators. The club-tipped tail of *Ankylosaurus* was probably a dangerous weapon.**

How did *Ankylosaurus* move?

Walking with Dinosaurs shows *Ankylosaurus* as a heavyweight, slow-moving herbivore that, like all dinosaurs, moves with its limbs held directly underneath its body and with its tail held stiff and well off the ground. Though some old illustrations show ankylosaurs walking with sprawling legs, complete skeletons show that such positions for the limbs were not possible, and the animals actually walked with their hands and feet directly underneath their bodies. This is further confirmed by ankylosaur tracks from the Lower Cretaceous of British Columbia and the Late Jurassic of England. These tracks show that the animals walked at moderate speed, and also that the hands and feet were more or less kept underneath the body.

Old reconstructions also sometimes show the ankylosaur tail dragging along the ground, in contrast to what is shown in *Walking with Dinosaurs*. As is true of sauropod tracks and those of most other dinosaurs, ankylosaur tracks do not show drag marks made by the tail, suggesting that the tail was held up off the ground. The way the bones fit together in well-preserved skeletons confirms this and also shows that the tail was carried virtually horizontally. Interlocking bony projections and ligaments enabled it to be held in this way with the minimum of muscular effort, and those ankylosaurs that, like *Ankylosaurus*, had enlarged tail clubs also had notably enlarged reinforcing bony prongs near the tail tip.

How did *Ankylosaurus* use its tail club?

In *Walking with Dinosaurs*, *Ankylosaurus* wields its large tail club as a threatening weapon and uses it to strike a *Tyrannosaurus*. It appears likely that *Ankylosaurus* used its tail club in this way, and indeed there is possible evidence that it did. However, this subject has recently been controversial. The traditional assumption is that the tail was mobile enough for these animals to have thrown their tail clubs around with vigor, but some ankylosaur experts question this. In particular Dr. Walter Coombs has argued that the ankylosaur tail seems to have been too stiff to have allowed much side to side movement, and that using the club to fend off attackers seems unlikely. Perhaps it was used instead as a sexual display symbol, or in ritualized combat with other ankylosaurs.

Other ankylosaur experts think that Coombs is wrong, and that the long ribs that grow sideways from the vertebrae near the base of the tail show that huge muscles were attached here. Therefore, the animals did have the capacity for powerful tail movement and only the end part of the tail was stiffened and relatively immobile. This would actually have been beneficial, as the stiff part would have acted as a sort of bony "handle" for the large club. Possible evidence that ankylosaurs could, and did, use their clubs in self-defence comes from studies on tyrannosaur bones. Several individuals are preserved with broken ankles, and of course "Sue" has a broken shin bone. These might be injuries that resulted from encounters with *Ankylosaurus*!

Torosaurus and *Triceratops*: horned giants from the Late Cretaceous

A second group of quadrupedal herbivorous dinosaurs, the horned dinosaurs or ceratopians, was also important in this Late Cretaceous world, and two types star in *Walking with Dinosaurs.* While all of the early horned dinosaurs had been sheep-sized or smaller, Late Cretaceous forms were mostly as big as modern rhinos and elephants. *Triceratops* and *Torosaurus*, both from the latest Cretaceous of North America, were among the very biggest, reaching 30 feet in length and perhaps 11 tons in weight. As seen in *Walking with Dinosaurs, Torosaurus* is remarkable for the size of its frill, and the biggest individuals have skulls ten feet long! Among land animals, only other kinds of horned dinosaur had skulls bigger than this.

Did *Torosaurus* live in herds?

Walking with Dinosaurs shows *Torosaurus* moving in herds, and these herds included juvenile animals as well as adults. Until very recently, there was no evidence for this since *Torosaurus* is poorly known and represented by few individuals. In contrast, some other kinds of seem to have perished together, perhaps by drowning, when on migration. The resulting collections of dead animals, preserved as bonebeds that may be square miles in extent, include individuals from all parts of the population, from babies to old adults. The presence of babies in these herds suggests that they were looked after by the adults.

A *Triceratops* bonebed has never been discovered, and even though hundreds of complete skulls of this dinosaur are known, specimens are rarely found together. This suggests that *Triceratops* was not a herding dinosaur. Evidence that *Torosaurus* did move in herds, as shown in *Walking with Dinosaurs*, came in 1999 when the discovery of a *Torosaurus* bonebed was announced.

What did baby torosaurs look like?

No juvenile *Torosaurus* specimens are known, but in *Walking with Dinosaurs* we see juveniles that are equipped with shorter frills and horns than adults. What is the basis for this reconstruction, if the fossils of babies are unknown? Baby specimens are known for other horned dinosaurs, including for *Triceratops*. As is true for most baby animals, the skulls of these specimens are proportionally large and their faces are shorter and their eyes bigger. The frills of the babies

▽ **Horned dinosaurs like *Torosaurus* had the biggest skulls of any land animals ever. Their huge frills and prominent horns were probably used in display, and certainly in battle.**

are not as proportionally large as in adults, and their horns are only present as tiny stumps. Using this information, *Walking with Dinosaurs* could reconstruct juvenile specimens of *Torosaurus*.

What did *Torosaurus* use its frill for?

Torosaurus was depicted in *Walking with Dinosaurs* with a striking, brightly colored "false eye" pattern on its frill. It is shown using this frill as a social signal in disputes with other males, and also fighting with its horns when this display fails to impress its opponent.

Obviously there is no direct fossil evidence for the "false eye" pattern, but it does seem likely that horned dinosaurs used their frills in this way, both to intimidate predators and enemies, and to appear attractive during courtship. Evidence that the frills were used in courtship comes from studies on dimorphism in these dinosaurs – many species exhibit two somewhat different forms, reasonably inferred to be males and females.

As shown by Dr. Thomas Lehman in his studies on horned dinosaurs, individuals with relatively straight, parallel horns seem to be males, while those with inclined horns that diverge away from one another seem to be females. There also appears to be some variation in frill shape between the sexes. This has been analyzed in the early horned dinosaur *Protoceratops* and some individuals, presumably the males, have larger, more erect frills. It also appears that male horned dinosaurs were bigger than females.

△ *Triceratops* is the most famous of the horned dinosaurs, and one of the largest. Unlike most horned dinosaurs, *Triceratops* had a solid frill that lacked perforating "windows".

Did torosaurs fight with one another?

When visual display among the torosaurs in *Walking with Dinosaurs* fails, the males engage in combat, locking their horns together and engaging in a twisting and shoving match. Various evidence allows paleontologists to be confident that horned dinosaurs did behave like this. Firstly, that male horned dinosaurs had different horns and frills from females suggests they were using them differently, and display and combat seem the obvious purposes. Many paleontologists have compared the horns of these dinosaurs with those of living antelopes and sheep, animals in which the males fight with one another for breeding rights.

Horned dinosaurs do not appear to have been particularly aggressive animals, as in contrast to species like *Tyrannosaurus rex*, the number of broken and damaged bones seen is quite low. However, some specimens do exhibit facial injuries that must have been caused by other horned dinosaurs. One *Triceratops* skull has a hole through its cheek region, perhaps caused by the horn of another *Triceratops*, and specimens of *Torosaurus* and other types are also known with such holes. Some horned dinosaur skulls have missing horn tips that may have broken off during combat.

Prior to this interpretation of the frills and horns as courtship and combat structures, the huge frills of horned dinosaurs were thought to have evolved either as defensive structures, or as attachment sites for huge, elongated jaw muscles. Both ideas are now deemed unlikely. It does not seem that the frills were for defence as nearly all kinds of horned dinosaur had large, vulnerable openings in the frill. *Torosaurus* had these openings but *Triceratops*, which lacks them, was one of a handful of exceptions.

As for the muscle attachment argument, nearly the whole frill surface lacks the distinctive surface texture made when muscles attach to bone. Horned dinosaur expert Dr. Peter Dodson has also pointed out that these dinosaurs would not have needed jaw muscles this long (more than a yard long given the size of the frill in *Torosaurus* and other forms). No living animal, even those that bite and chew very tough vegetation, has jaw muscles even approaching three feet in length.

Hell Creek's "giant duck"

Another large dinosaurian herbivore that features in *Death of a Dynasty* is *Anatotitan*, a hadrosaur with a ducklike snout. The animal's name commemorates this similarity, meaning "giant duck". *Anatotitan* was indeed a gigantic beast, reaching 43 feet in length. Among hadrosaurs it was peculiar in that it had a very long, low head with an especially elongate toothless beak at the front. The hundreds of tightly packed teeth and powerful jaws of these dinosaurs show that they did not live like ducks, but were instead well equipped to crop and chew tough vegetation.

Named by Dr. Michael Brett-Surman in 1990, *Anatotitan* was discovered in the nineteenth century but had been thought to be a species of the similar, closely related hadrosaur *Edmontosaurus*. Both *Anatotitan* and *Edmontosaurus* were flat-headed hadrosaurs, rather conservative in appearance compared with the crested hadrosaurs or lambeosaurs, a group that had been abundant earlier on in the Late Cretaceous.

The lifestyle and senses of *Anatotitan*

Walking with Dinosaurs shows *Anatotitan* as a herd-dwelling, terrestrial, quadrupedal dinosaur that keeps in contact with its herd members by making various calls. This view of hadrosaurs reflects a modern consensus that contrasts with the nineteenth-century view.

Because their tails are deep and their beaks superficially like those of ducks, nineteenth-century paleontologists assumed that hadrosaurs were amphibious. Their teeth were thought to have been weak and best suited for chewing soft water plants. This seems ridiculous given the hundreds of closely fitting, rasping teeth in the hadrosaur jaw, but it became entrenched in the literature. It is now clear that hadrosaurs were not adapted for an amphibious life – their tails were stiff and their hands were not paddle-shaped, as is shown by complete skeletons and fossil tracks. Hadrosaurs have the same kind of chewing mechanism as seen in *Iguanodon*, and their magazines of interlocking teeth appear suited for chewing coarse plant material.

The system of calls the *Anatotitan* make in *Walking with Dinosaurs* is inspired by paleontological work on the nasal regions of these dinosaurs. *Anatotitan* and related hadrosaurs have enormous nostrils surrounded by a hollowed out area on the snout. These defy explanation, but one idea is that they housed inflatable skin sacs that could have been used to improve the resonance of their calls, and might also have served as visual display signals. The large spaces for eyes and the delicate ear bones in hadrosaurs show that their eyesight and hearing were acute, and the large nostrils may also show that they had a good sense of smell and perhaps used this to detect approaching predators.

Dinosaur-killing alligators

Tyrannosaurus was the biggest Late Cretaceous predator on land, but equally formidable reptiles lurked in the waters of the time. *Walking with Dinosaurs* shows the immense alligator *Deinosuchus* attempting to ambush the pterosaur *Quetzalcoatlus*. While most of the crocodiles and alligators of dinosaur times were not particularly big, usually being less than ten feet in length, *Deinosuchus* was exceptional, possibly reaching 33 feet and more. It seems obvious that *Deinosuchus* grew to such sizes so that it could ambush and kill large dinosaurs. Evidence that it did so comes from the bones of an *Albertosaurus*, a relative of *Tyrannosaurus*, that have *Deinosuchus* tooth marks on them.

How *Deinosuchus* grew to be so big has recently been the subject of investigation. Studies of the microscopic structure of their bones show that dinosaurs reached their giant size by growing very rapidly, more rapidly than is possible in crocodiles and alligators. Was *Deinosuchus* also capable of particularly rapid growth? Dr. Gregory Erickson and Dr. Chris Brochu showed, by examining cross-sections of *Deinosuchus* bones and armor plates, that it adopted a different strategy. Whereas none of the related forms of alligator seem to have had lifespans exceeding about 25 years, and all increased markedly

◁ **Duck-billed dinosaurs, like these *Anatotitan*, were common herbivores in the Late Cretaceous. They were well suited for eating tough vegetation.**

◁ ***Deinosuchus* was a huge alligator, perhaps more than 33 feet in length. It is known to have eaten dinosaurs.**

in size during their first decade only, *Deinosuchus* lived to over 50 years and maintained a juvenile-style growth rate for several decades. While *Deinosuchus* did rival giant dinosaurs in size, it did so by employing a different, much slower style of growth.

Rightly named to mean "fearfully great crocodile," *Deinosuchus* is best known from a full-sized reconstruction of its immense skull displayed at the American Museum of Natural History in New York. In contrast to this skull, which is notable for its massive, tall snout tip, the *Deinosuchus* in *Walking with Dinosaurs* has a shallow snout tip. However, the New York skull was reconstructed on the assumption that *Deinosuchus* was a crocodile. As it is now clear from special features in the jaws that it is an alligator, its snout should have been lower.

At present, *Deinosuchus* is not known with certainty from the Hell Creek, coming instead from the older Cretaceous rocks of Montana, Texas, Georgia, and elsewhere. It does not therefore seem to have lived alongside *Tyrannosaurus* and *Triceratops*.

The first snakes

Walking with Dinosaurs also showed another modern looking reptile that lived alongside the Late Cretaceous dinosaurs, a constricting snake. Snakes were clearly present by Late Cretaceous times and modern families, like boas, were present by Hell Creek times. Vertebrae from a boa have in fact been found in the Hell Creek Formation, though of course this would not have been the modern-day species shown in *Walking with Dinosaurs*!

In *Walking with Dinosaurs*, the snake is shown as having heat-sensitive pits in its snout that allow it to detect the presence of warm-bodied prey animals. Such pits are seen in three groups of snakes today: boas, pythons, and vipers. The distribution of pits in the members of these groups show that the structures evolved independently each time, and were not inherited from earlier snakes.

The controversial origin of snakes

While it is universally agreed that snakes are part of the same group of reptiles as lizards, called squamates, experts disagree over the exact relationship of the two groups and over the evolutionary transition that resulted in the appearance of the very earliest snakes.

One school of thought is that snakes arose from advanced lizards related to monitor lizards and the sea-going mosasaurs, and like mosasaurs, the very first snakes were marine. Backing for this theory comes from a variety of Cretaceous snake fossils which appear to be both primitive relative to other snakes (they still have tiny hindlimbs for example) and were clearly marine (their stomach contents show that they ate marine fish). Those who support this view argue that it was only later on that snakes took to life on land.

An opposing school argues instead that snakes descend from very primitive lizards, or perhaps not from lizards at all, and that the most primitive snakes were terrestrial burrowing forms. According to this theory, the Cretaceous marine snakes are actually advanced, rather than primitive, and the similarities seen between snakes and the lizards like mosasaurs are the result of similar feeding behaviors.

Quetzalcoatlus, last of the pterosaurs

Throughout the *Walking with Dinosaurs* series, a diverse assortment of pterosaur types were shown. Now, at the very end of the Cretaceous, only one or two kinds were present. *Death of a Dynasty* shows *Quetzalcoatlus*, an enormous gliding form discovered in Texas in 1975. Though it was originally estimated to have a wingspan of around 50 feet, most current estimates put the figure for *Quetzalcoatlus* at around 36 feet. Little information about its skeleton is known, and determining the shape of its skull has been problematic. *Walking with Dinosaurs* gave *Quetzalcoatlus* an elongated bill with a short, backwards-pointing crest on the top of its head. Partial skulls of small *Quetzalcoatlus* individuals show a different shape from this, with a raised ridgelike crest in front of the eyes. However, these small individuals may not be of the same species as the giant, 36-foot specimen, so its skull shape remains mysterious.

Walking with Dinosaurs shows *Quetzalcoatlus* feeding by plucking a fish from the surface of a lake. It is not really known how *Quetzalcoatlus* and its relatives fed, and there have been suggestions that it might have fed on dinosaur carcasses or on burrowing molluscs and crustaceans. These do not seem possible given the straight, rather

THE RISE OF BIRDS

WHILE SOME LATE Cretaceous dinosaurs had become the giants of their respective groups, others had become small. Birds, in particular, appear to have become miniaturized since evolving from their theropod ancestors. Whereas an "average" theropod might have been the size of a wolf, if not bigger, an "average" Cretaceous bird would be the size of a modern starling.

We do know, from Late Cretaceous fossils from North and South America, Mongolia, and China, that birds had become important as tree-dwelling predators of insects and other small animals, and were also abundant on shorelines and mudflats. Late Cretaceous bird fossils from New Jersey include forms rather like modern sandpipers and curlews as well as possible relatives of albatrosses. Numerous fossil tracks from the Early Cretaceous of North America, Korea, and elsewhere show that such waterbirds had been abundant and diverse since even earlier times in the Cretaceous. More specialized waterbirds that were now flightless had also evolved by the Late Cretaceous and hunted fishes alongside the marine reptiles of the time. Flightless running birds, something like living rheas and ostriches, were also now present and there is even evidence that the first members of some rather more modern groups, including parrots, were present in the Late Cretaceous.

Perhaps the diversification of birds in the Cretaceous partly explains the decline in the other flying archosaurs, the pterosaurs. Known Cretaceous pterosaurs appear to have been almost entirely restricted to aquatic environments, in contrast to birds which clearly evolved to exploit terrestrial habitats like woodlands. Seeing as birds were more adept at moving on their hind legs than pterosaurs, it seems that birds replaced pterosaurs as the birds moved into waterside habitats. This would have occurred slowly over the Cretaceous as birds gradually became more widespread, not because birds and pterosaurs actively competed with one another.

▷ **With a wingspan of around 36 feets, *Quetzalcoatlus* was one of the largest, and last, of the pterosaurs.**

inflexible neck of this pterosaur and the shape of its huge bill. It therefore seems likely that *Quetzalcoatlus* fed by picking prey up from the water or from the ground.

Didelphodon: the "Tasmanian devil" of its day

Walking with Dinosaurs focuses on one Late Cretaceous mammal, the boldly patterned scavenger *Didelphodon*, and shows it opportunistically raiding dinosaur nests and scavenging from dinosaur carcasses. We know little about the ecology and behavior of this animal (and, needless to say, we know nothing of its color scheme (see the introductory chapter), but its appearance symbolizes increasing diversity in Late Cretaceous mammals.

Didelphodon was a stagodontid, a group of mammals belonging to a branch called the metatherians. Metatherians had first appeared in the Early Cretaceous, probably in North America, and later spread into the Southern Hemisphere where they evolved into marsupials, the modern pouch-bearing mammals. Being about the size of a badger, *Didelphodon* was one of the biggest mammals known for the whole of the Mesozoic. The Cretaceous platypus *Steropodon* from Australia was of a similar size.

Was *Didelphodon* a scavenger and egg eater?

Walking with Dinosaurs shows *Didelphodon* eating the carcass of a young *Torosaurus*. Such behavior would have been unusual for most Mesozoic mammals – they were mouse-sized predators of insects and other small animals, unequipped to eat dinosaur bones and skin. However, the lower jaw of *Didelphodon* allows us to make some inferences about the diet and lifestyle of this animal. Its jaw is deep and powerful and its canine teeth are thick and conical – features

▽ By Late Cretaceous times, small mammals were diverse and abundant. *Didelphodon,* a distant relative of modern marsupials, was one of the biggest Cretaceous mammals and was probably an opportunistic predator and scavenger.

showing that it had a powerful bite and was well equipped to bite large objects. Its premolar teeth are bulbous and powerful and have been compared to those of the Tasmanian devil, a living marsupial that eats carcasses and can break open bones. Based on this analogy, *Didelphodon* might also have been a scavenger and a veritable Tasmanian devil of its day.

Didelphodon in *Walking with Dinosaurs* is a largely solitary animal that fights and squabbles with other members of its kind. We do not really know that it behaved like this, but it seems a reasonable inference. Living carnivorous marsupials, the closest living relatives of the stagodontids, are solitary predators that react aggressively when they encounter members of their own kind. When Tasmanian devils meet, for example, they hiss, growl, and strike at each other until the subordinate individual backs down. It is tempting to suggest that *Didelphodon* might have behaved similarly.

Walking with Dinosaurs also shows *Didelphodon* stealing eggs from the nest mound of a *Tyrannosaurus*. It is possible that *Didelphodon* exploited food sources such as these, much as most modern carnivores will steal and eat bird's eggs if they find them. However, its teeth and jaws show that it was clearly not well equipped to break into and eat eggs.

The end-Cretaceous extinction event and the death of a dynasty

As the title suggests, *Death of a Dynasty* ends with the catastrophic extinction event that brought an end to the reign of the dinosaurs. However, *Walking with Dinosaurs* depicts the dinosaurs as already in decline due to an increasingly volcanic atmosphere.

This decline presages the impact of a huge comet at the Yucatan Peninsula, Mexico. The impact of this comet sends millions of tons of rocky debris into the atmosphere; the heat generated by the impact starts wildfires on a global scale, and the impact itself sends out an immense shock wave that generates huge tsunamis (tidal waves). Already reduced in diversity and struggling in a polluted world, the dinosaurs in *Walking with Dinosaurs* vanish from the Earth forever.

Evidence for an extraterrestrial impact

The idea of a comet or meteor hitting the Earth at the close of the Cretaceous was not really thought scientific until 1980. In that year, Dr. Luis Alvarez and his colleagues reported the discovery of high concentrations of iridium in latest Cretaceous strata in Italy. This iridium layer was later found worldwide. Iridium is rare on Earth and mostly reaches the planet via meteors. Presumably, therefore, the Late Cretaceous iridium had an extraterrestrial source, and most reasonably must have entered the Earth's atmosphere as a meteor or comet.

If this happened where was the impact crater? The recognition of such a crater, formed in rocks of exactly the right age to coincide with the extinction, was made in 1990. An immense circular structure, now called the Chicxulub Crater, was discovered on the seafloor just off and under the Yucatan Peninsula, Mexico. Subsequent studies revealed that small glassy fragments and pieces of quartz with "shock" marks were present in the rocks ringing the Chicxulub site. The glassy fragments had clearly been created by intense temperatures, and the shocked quartz was evidence of a devastating collision force. Beds of charcoal that resulted from wildfires have been found in the rocks around the Chicxulub site and jumbled piles of debris are preserved in strata that were well inland from the impact site – they seem to result from huge tidal waves created by the collision.

Were dinosaurs in decline anyway?

There is no doubt that a huge body did hit the Earth at Chicxulub at the very end of the Cretaceous. But was this event the single cause of dinosaur extinction? Extraterrestrial bodies appear to have hit the Earth during other parts of the dinosaurs' reign, and yet none of these events caused their total extinction. Furthermore, some kinds of dinosaur seem to have been dwindling prior to the very end of the Cretaceous.

One theory is that volcanic activity was increasing in the Late Cretaceous, and the resulting acidification of the water was creating a decline in dinosaur and other animal populations. It has also been pointed out that certain products of volcanic activity prevent reptile eggs from forming properly. Large amounts of these products would have resulted in fewer successful hatchings, and therefore a gradual decline in populations. However, the evidence for increasing volcanicity is controversial, and strongly opposing views are expressed by different scientists. Some maintain that there was intense volcanic activity in the Late Cretaceous. Others contend the opposite: that the Late Cretaceous was, from a volcanic point of view, a "quiet" period.

Though at least some Late Cretaceous dinosaur groups were extinct before the very end of the period, were all dinosaurs really in decline? This is another area of much controversy. The latest Cretaceous rocks of North America appear to show that only a handful of dinosaurs were still abundant, among them *Triceratops* and *Edmontosaurus.* This decline is attributed by some to the retreat of coastlines which would have resulted in the loss of moist coastal environments in which many Late Cretaceous dinosaurs had flourished. The remaining low number of species would have made the group as a whole vulnerable to extinction as the potential for the evolution of new species would have been lower than before. However, was dinosaur diversity in the latest Cretaceous really this low? Some pale-

▷ **Some experts argue that volcanic events at the end of the Cretaceous created a more hostile climate, and that this caused dinosaur extinction.**

ontologists argue that it was not and point to records of small theropods and other dinosaurs from the latest Cretaceous.

If anything has become clear from investigations on the last dinosaurs, it is that extinction events are not simple, and that it is difficult to extrapolate clear patterns from the many lines of evidence. It remains mysterious why dinosaurs and other animals became extinct, while others, including amphibians and the crocodile-like champsosaurs, survived unscathed. Perhaps the greatest problem in the debate about dinosaur extinction is that virtually all of our data comes from western North America, and what was happening here may not have been true of the rest of the world. While it now seems that the impact of an extraterrestrial body was the final blow that finished off the dinosaurs forever, it also appears that changes in the environment were resulting in their demise.

The end of an era

Walking with Dinosaurs depicted the Late Cretaceous world as a time of turmoil. The forces of nature were inadvertently conspiring against the dinosaurs, and were soon to put an end to their time on the Earth. This gradual decline remains controversial, and some paleontologists think that dinosaurs were still flourishing at high diversity up to the final impact event. Even if this were so, evidence for environmental change indicates that there was less suitable habitat space for dinosaurs than there had been before. *Walking with Dinosaurs* also reconstructs the impact event itself, as animals living well north of the Mexican impact site would have experienced it. This would have been a truly terrifying event.

At the impact site, and within a few hundred to thousand miles of it, dinosaurs and indeed most plant and animal life was killed off instantaneously. Beyond this area life was affected in other ways. Huge forest fires swept the landscape and would have wiped out millions of species from lowly invertebrates to forest-dwelling birds and dinosaurs. Clouds of dust would have choked animals where it fell back to Earth and plants would have been buried concealing the food of herbivorous animals.

Where fallout was less intense changes occurred more slowly. Climatic changes would have wrought havoc to populations of animals adapted to warm tropical regimes when the temperature plummeted, while changes in rainfall pattern would have devastated arid or desert environments.

The long periods of darkness that fell when the sun was obscured by the dust reaching higher levels of the atmosphere would have caused plant life to wilt and die, although funghi may have fared better. Their recovery may have been too slow for many of the larger herbivores and gradually the food chain would have collapsed – beginning with the plants and followed by the herbivores starving to death. Predators were hard hit too, for once the largest herbivores had gone, giants like *Tyrannosaurus* would have starved to death. Scavengers would probably have done well at this time, perhaps for several months or even years in some places. In a world awash with cockroaches and flies, insectivores would have seen good times. Competition for food among all the animals, though, would have been intense and eventually populations of all of the large terrestrial animals would have dwindled to unsustainable levels – within perhaps a few decades the large dinosaurs would have disappeared completely.

Perhaps changes in temperature sent crocodiles into a sort of hibernation; small furry mammals may also have been able to take refuge, but birds were not able to hibernate. Perhaps the highly mobile birds were able to seek out safer places and range more widely in search of food. Perhaps some birds fared better during this time. With an increased supply of insects, which were feasting on the rotting carcasses that must have littered the Earth, birds did not go hungry.

After the final extinction of the dinosaurs and the last pterosaurs, the Earth must have been a fairly empty place, devoid of all animals larger than about ten feet. New opportunities now presented themselves to the small animals that inherited this world and, with dinosaurs out of the way, mammals became gradually larger and increasingly more diverse. In effect, we probably owe the rise of our own group, and that of all the major groups of animals around us, to the extinction of the dinosaurs.

The impact theory itself seems to be well established. The really important area of interest to paleontologists now is in trying to unravel the details of the aftermath, and not just for interest's sake, but because there is the very real possibility that the Earth will suffer another devastating impact some time in the future.

Index

Numbers in *italics* refer to illustrations; numbers in **bold** refer to boxes